RIGHT BACKED BY MIGHT

RIGHT BACKED BY MIGHT

The International Air Force Concept

ROGER BEAUMONT

PRAEGER

Westport, Connecticut
London

Library of Congress Cataloging-in-Publication Data

Beaumont, Roger A.
Right backed by might : the international air force concept / Roger Beaumont.
p. cm.
Includes bibliographical references and index.
ISBN 0-275-97172-4 (alk. paper)
1. International police. 2. Air forces. I. Title.
JZ6374.B43 2001
327.1'72—dc21 00-058023

British Library Cataloguing in Publication Data is available.

Library of Congress Catalog Card Number: 00-058023
ISBN: 0-275-97172-4

First published in 2001

Praeger Publishers, 88 Post Road West, Westport, CT 06881
An imprint of Greenwood Publishing Group, Inc.
www.praeger.com

Printed in the United States of America

The paper used in this book complies with the Permanent Paper Standard issued by the National Information Standards Organization (Z39.48-1984).

10 9 8 7 6 5 4 3 2 1

To my children

Anne Eric Katherine

Gemma sunt

Contents

Introduction ix

I. Seeds on the Wind: Visions of Universal Peace and Air Power 1

II. Bloom and Wilt: The International Police Force Concept and the League of Nations 19

III. Blossoms on the Wind: Growth of the International Air Force Concept, 1919–1939 33

IV. In Fullest Bloom: Manifold Visions of an International Air Force in World War II 71

V. Blossoms Scattered and Faded: Collective Security since World War II 123

Index 173

Introduction

While this study of the International Air Force (IAF) concept began in earnest during a one-year stint at the Naval Academy 1989–1990, my interest had been piqued over the years by passing references to the IAF in Air Ministry Records in the Public Record Office (PRO), and U.S. Air Force histories. That, and vague memories from my childhood of fanciful images of United Nations (UN) armed forces keeping the peace led me to assume that the subject was obscure. I anticipated finding enough material for an article, with perhaps two or three dozen footnotes but I soon found that the International Air Force idea and the closely linked International Police Force (IPF) concept were ideas that have come in and out of focus in popular culture and public policy process and analysis throughout the twentieth century.

Many people helped me follow the various threads of this intricate tapestry, including the staffs of: the U.S. Historical Center at Maxwell Air Force Base, Alabama; the New York Public Library; the U.S. Naval Academy Library; the British Public Record Office; the Rice University Library; and at University of Texas-Austin's Perry-Castaneda, Undergraduate, and Humanities Research Center Libraries. I am especially indebted to the interlibrary loan staff at the Sterling C. Evans Library at Texas A&M University, and very grateful for materials by General Lucien Robineau and the research staff of the Service Historique de l'Armée de l'Air at Vincennes and for information offered by Colonel Ron Ladnier of the Politico-Military Affairs Bureau, U.S. Department of State. Beyond that, special thanks are due my mentor and old friend Robin

Higham, Texas A&M University History department heads Larry Hill and Julia Kirk Blackwelder for their assistance and encouragement, and Texas A&M University for a Faculty Development leave that allowed me to conduct research in Britain. My daughters Anne and Katherine, and son Eric, offered a patient ear from time to time, as did my colleague John Robertson, while my wife Penny provided vital and steadfast aid in research and editing.

I

Seeds on the Wind: Visions of Universal Peace and Air Power

Over the last half millennium, dozens of plans have been devised for maintaining or imposing peace by multinational armed forces or associations of nations. While such schemes were long judged as fanciful or impractical, the tide seemed to be changing from the mid-1800s onward, for several reasons. The great powers were involved in the last phase of imperialist expansion, imposing by force of arms, or the threat of it, what many—other than those caught in the direct path of those efforts—saw as a more civilized and tranquil order based on and along with common identity and law. The industrialized West, or what many today call the North, held a strong edge in technology that gave them a substantial edge over what Kipling, as a bard of imperialism, deemed "lesser breeds without the law." The "powers-that-be" in that age were widely—but not universally—perceived as having lofty goals and values, which along with basic greed and search for dominance drove their expansion across the world from the fifteenth century onward. Apologists and enthusiasts of modern imperialism often invoked ancient models, especially Rome, which had used force to impose a relatively peaceful order, to justify the Pax Britannica, as well as France's civilizing mission, American "manifest destiny," and ultimately, Italian Fascism and Nazism.

From the Franco-Prussian War of 1870–1871 to the outbreak of World War I in 1914 a widely shared sense that the world was evolving toward a more peaceful condition was bolstered by an interval of tranquility and abundance in western and central Europe, which contrasted with a chronic flickering of colonial "small wars" and troubles in such unsettled areas as the Balkans and Ireland. Other bright spots were the growing willingness of nations to discuss arms limitations, however slight and symbolic the agreements were, and their submitting disagreements to arbitration, albeit in matters far from their vital interests. Those trends gave fresh impetus to the old visions of an international police force, which enthusiasts traced far back into history to such examples as the Amphictyonic League of classical Greece, which had required members to submit "disputes to arbitration before resorting to war."[1] On an intellectually loftier plane were the proposals of Zeno, in the third century B.C. and Epictetus, two centuries later for creating systems of "world" order in which parochial differences would be resolved by reasonable accommodation. Many other such concepts emerged over the centuries before the advent of modern nationalism, all to no effect. One such model was designed by a French lawyer, Pierre du Bois in the early fourteenth century. Arguing that the Crusades, for all their horrors, had demonstrated the value of international military cooperation, du Bois proposed a Christian federation under French dominance, with a council, a tribunal, and mechanisms for appealing cases to the Pope, which would employ "concerted military action against . . . [an] offending nation."[2] A century later, in *De Monarchia*, Dante crafted a plan for a worldwide super empire with the power to bring bellicose states to heel.

While sovereignty was usually recognized as the nemesis of such schemes, clerical-secular struggles also thwarted the best intentions of some leaders who were willing to trade away some of their suzerainty for greater mutual security. In talks at Venice in 1462, for example, Papal diplomats blocked the framing of a formal agreement devised by King George Poděbrad of Bohemia's chancellor for forming a league of several major states against the Turks, including a congress of ambassadors.[3] Whether or not that was really a lost opportunity or merely a case of *realpolitik*, higher barriers to international association were erected throughout the early modern era by surging nationalism and imperialism in western Europe and which produced an ongoing barrage of hopeful treatises, preventing any local league that actually took form from evolving into a more general system of order. Ironically, the trend toward the centralization of power at the end of the fifteenth century yielded some benefits when internal feudal warfare was being stifled by the emergent nation-states. A century later, war among nations grew in scale and horror, and became more and more a means of resolving violence between sovereign powers, and less and less a way to resolve local squabbles and grudges.

Amid that transition, the most famous plan for regulating the new order of conflict appeared—the *Grand dessein*, described in the memoirs of Henri IV of France's minister, the Duc de Sully, as the product of his king and Elizabeth I. Less grandiose than earlier models, it was ostensibly aimed thwarting the Habsburg emperor and the Spanish king Philip II's ambition to unify Christendom against such infidels as the Dutch, English and Turks. Whomever its authors may have been, the *Grand dessein* proposed dividing Europe into fifteen states of nearly equal size, religiously balanced, with international land and sea forces funded by members proportional to their wealth. The main strategic goals of the scheme, aside from thwarting Spanish hegemony, were the conquest of Asia and Africa and the maintenance of peace. It defined "the great and common interest of Europe" as being of so great a good that "the greater powers should force the lesser into it, if necessary, by assisting the weak and the oppressed; this the only use they ought to make of their superiority."[4] Whatever Henri's—and perhaps Elizabeth's—purposes were after Philip's fortunes peaked in the naval victory over the Turks at Lepanto in 1571, the latter's hopes crashed with the defeat of the Armada in 1588 and the rise of the Dutch Republic. Amid the subsequent turmoil, the *Grand dessein* remained in view as a prototype, but never more than the most visible model of many such plans and ultimately no more tangible than the rest. In a very similar vein, for example, in the early seventeenth century, a Spanish theologian wove an elegant tapestry of ecclesiastical logic to support his plan for federation based on the premise that nations were members of a broader community. In 1623, Emeric Crucée in *Nouveau Cynée* set forth the first plan for forming a worldwide organization, this one to be headed by a Grand Council in Venice, with slightly different goals: keeping the peace, and halting dangerous trends toward democracy. Crucée's scheme also highlighted another stumbling block that would hinder many other attempts to build an international order based on force—the fear of a superorder imposing its will on a particular locality.

Many designs for a multinational order that appeared in the late sixteenth century were reactions to the horrific devastation of the Thirty Years' War in Germany. Most significant was the proposal of Hugo Grotius, later deemed the father of modern international law, for an international tribunal, as well as his treatise *De Jure Belli ac Pacis,* which traced precedents for forming an association of sovereign powers to scripture and canon law. While such blueprints during that era were generally aimed at restoring Europe's balance of power, or what Sully had called "equilibrium," echoed earlier designs, they also reflected that era's orderly mechanistic thought as exemplified by Newton, Napier, and Pascal. That rising strain of mechanistic rationality was visible at the end of the century in William Penn's essay "Towards the Present and Future Peace of Europe, by the Establishment of a European Diet, Parliament or Estates." (Both he and fellow Quaker John Bellers diverged from their brethren's pacifistic ideal in suggesting the use of force in their respective peace enforcement models.) Sully's theme of an international army was visible in Penn and Bellers' works and in another landmark treatise of those times, Abbé de St. Pierre's *Perpetual Peace (Projet de Paix Perpetuelle en Europe)*, published in 1713. Following his service as secretary to a French representative during the drafting of the Treaty of Utrecht, which ended the War of the Spanish Succession, St. Pierre set forth rules for an association of states using their combined forces against any power that initiated a war, violated a regulation or judgment, refused to join, committed aggression, launched a sudden attack, or defied the association. He also offered provisions for changing boundaries peacefully. St. Pierre's treatise was widely circulated, and threads of it would appear in peace enforcement proposals over the next three centuries,[5] including the American colonists' Albany Plan for creating a mutual security association in 1745, Jean-Jacques Rousseau's essay "Lasting Peace Through the Federation of Europe" in the 1750s, and Immanuel Kant's *Thoughts on Perpetual Peace* (1795).

Early in the nineteenth century, as the great European powers squabbled over various designs for an association to maintain the peace in the wake of the Napoleonic Wars, they encountered the old dilemma: in creating such a union and forming an international police force, whose law and which order would be maintained and enforced? That point, also made by the Soviet Union in response to Spanish and French proposals for an international air force during disarmament talks at Geneva in 1932, had come to the forefront at the Congress of Vienna in 1815. There Czar Alexander I, in the manner of a chess strategist, offered to station 150,000 Russian troops in northern Italy as a check against revolution and political ferment. When the major continental powers formed the Concert of Europe, Britain demurred, and focused on its overseas empire, suppressing the transatlantic slave traffic and stirring up revolutions in Latin America that opened up the area to free trade—that is, commerce favorable to Britain's maintaining its lead in the Industrial Revolution.

As noted earlier, Pax Britannica proved more relative than absolute. Over the next generation, the spirit of liberalism stirred up by the French Revolution was smothered repeatedly by the Holy Alliance under the orchestration of the Austrian statesman Metternich. In that era of centralism and despotism, several St. Pierrean models appeared, again without practical consequence,[6] although other currents were flowing toward framing a peaceful international order. In Britain, for example, the energetic reformer Richard Cobden turned his attentions to disarmament after his successful crusade to repeal the Corn Laws, playing a key role in keeping Britain from going to war during "The Three Panics" of 1848, 1853, and 1862. While he favored some sort of international peace-enforcing mechanism, his distinguished former colleague, John Bright, held the contrary view that "no nation or group of nations, led by no matter whom, can force peace upon the world. Force is no remedy."[7] As an American peace movement gained momentum on a parallel track, the long series of Lake Mohonk conferences increased idealists' awareness of practical constraints while bolstering a sense of fellowship and fed an optimism that survived the Mexican and Civil wars. Although most American pacifists rejected using force, they favored creating an international tribunal and forms of what would later be labeled "binding arbitration" and "dispute resolution" in settling differences.

On another path, the growing destructiveness of military weaponry led an increasing number of European rulers and diplomats to seek to impose controls on armaments through multinational treaties and to submit international disputes to negotiation. Their concern mounted sharply in the 1850s, when the Crimean War showed the grisly effects of technological evolution on the waging of war. The bloody results of rifle fire in close combat at Solferino in 1859, which literally nauseated Napoleon III, led to the creation of the Red Cross. Despite the much greater slaughters in the American Civil War and Prussia's wars of expansion, the spirit of international accord seemed to grow in the late 1860s, when Prince Peter of Oldenburg visited Napoleon III, Alexander III, and the king of Prussia, proposing all civilized nations renounce war, an international arbitration committee be formed, and that an international convention be established to regulate all armed forces. Despite many nods and murmurings of approval, nothing substantive came of Peter's plan, but the horizon seemed to continue to brighten during the last half of the nineteenth century and early 1900s, as many nations used arbitration to work out disputes. While the series of international conventions that began at the urging of Czar Alexander II after the Crimean War seemed altruistic, it also served Russian national interests, since major disarmament would slow the western European technical advances that allowed Russia's numerically inferior foes to remain on its doorstep despite its massive peasant armies. At that time, as they would be over the next 150 years, some arms control designs were aimed clearly at altering the balance of power, while others were based on appeals to reason and humanitarianism. Theorists created articulated models aimed at equitable balance including elaborate formulae for apportioning armed force among nations on the basis of factors like geography, wealth, population, boundaries, etc. But even though some attempts made over the centuries to bridle Mars prevailed, in general, optimism far surpassed feasibility, and sovereignty thwarted the hopes of the more ambitious designers. That was the case, for example, with James Lorimer's 1867 proposal for mixing national constabularies and "a small international army . . . of contingents supplied by the separate member states" to keep the peace, which did not preclude such forces from maintaining the status quo within nations or empires.[8]

Yet, despite the misplaced enthusiasm and naiveté visible at the time or apparent in hindsight and all the floundering and futility, not all the efforts of "men of good will" during the nineteenth and early twentieth centuries came to naught. They were able to curtail some especially nasty types of weapons, like dum-dum and glass bullets, and 250 successful arbitrations took place from 1815–1914. Britain's indulgence of American feistiness in the Canadian boundary and *Alabama* claims disputes may have been due more to a search for allies as Britain's imperial power declined than to altruism, but many pacifists and others took heart from it, as they did from the smooth operation during wartime of such functional international agencies as the postal and telegraph unions. Sometimes major imperial powers also used their might to do such good works as suppressing the slave trade and dispatching punitive expeditions then seen as bringing civilization and Christianity to "benighted heathens." In searching for precedents of muscular peace maintenance, architects of peace enforcement schemes pointed to some of the more savory examples of "gunboat diplomacy," like the Barbary Wars, antipiracy campaigns, Navarino, and certain cases of port-forcing and blockades. Most—but not all—colonial wars were brief and tiny in scale and inflicted relatively few casualties on Western forces involved, especially after medicine and microbiology reduced the toll taken by tropical diseases. Often, the mere appearance of imperial military or naval forces ended violence and restored order, or what those wielding preponderant force defined as order, but lurid details of cases where matters led to violence were not always made public, like the behavior of some European contingents at Peking in 1900.

While most of Europe was warless for a generation, colonial warfare was more and more equated with the types of policing, based on the British model that emerged in the early nineteenth century. The full implications of the major powers' armament production and alliances were discounted by most observers and elites alike. There was also a division of opinion between "pure" and "hard-headed" pacifists over the optimal mechanism for peace keeping. Many versions of Sully's proposals appeared, like Victor Hugo and Charles Lemmonier's "United States of Europe," linked to a "multinational peacekeeping arm," and the plan set forth in 1894 by Raoul de la Gresseria, a French judge, who urged the United States and Europe to create an international army as a prelude to a program of disarmament that he hoped would lead to the forming of a world federation. Although his scheme was widely discussed, it left the key issue of boundary adjustment unaddressed and fell by the wayside when the main focus of delegates at peace conferences at The Hague shifted from disarmament to arbitration, including establishment of an international court for such deliberations.[9]

In a grand contradiction, as the new century dawned European powers and the United States were engaged in the last spasm of imperial expansion at the same time that peace advocates in many nations were invoking American federalism and the British dominions as salient precedents for establishing international citizenship, a worldwide judicial system, and "an international police." On the crest of the rising optimism that marked the *Belle Epoque,* those goals were described "as certainly not impossible," since many believed that "war's days are nearly numbered" and that the world was approaching a state of tranquility in which "force will play a very small part."[10] Such gross miscalculations arose from widespread belief in the inevitability of progress, which had become a kind of lay faith throughout the industrialized world. Dramatic technological advances, a sense of growing wealth fueled by major gold strikes, and an apparent trend toward democracy fed hopes that major conflicts would soon be obsolete. Apparently powerful peaceful trends on the eve of World War I led Norman Angell to assert in *The Great Illusion* that war was most unlikely, since the world was ruled by sensible elites who were able to read a balance sheet and therefore likely to let things get out of hand that might threaten their investments and vital interests. Angell was only one of a great many intelligent people who vastly underestimated the volatility of the witches' brew of nationalism, imperialism, and high technology. The pot was not on a low simmer, as it seemed in that lush, wan twilight of the old order, but about to boil over. Harry Emerson Fosdick caught the essence of it when he pointed out the "sheer paganism" of mass nationalism in which "the supreme object of devotion for multitudes, is the nation. In practical action they know no higher God."[11]

The Russo-Japanese War of 1904–1905 offered many glimpses of what would soon erupt in Europe. It was the first major conflict to be presented in photographs to millions of readers of newspapers and magazines throughout the world, and it was the first in which quick-firing long-range artillery, repeating rifles, and machine guns were used in large numbers on both sides. A host of journalists and foreign military officers took detailed notes and wrote long reports, but ultimately more distant journalist-intellectuals came closer to the mark in anticipating where those trends were leading. When war came in 1914, Angell was proven wrong, when H. G. Wells' and Ivan Bloch's visions of a vast conflagration and mechanized warfare were realized as havoc welled up on a scale that stunned the world.

In 1914, with the coming of World War I, the framework of hope erected so painfully for almost a century including Universal Peace Congresses, meetings at The Hague and Geneva, the International Peace Bureau, the Mohonk meetings, the Czar's rescript of 1898, dozens of arbitrations, and hundreds of treaties crumbled like matchsticks under a steamroller.[12] At the outset, military operations lunged out of the control of politicians, diplomats, generals, and admirals and gained a momentum and gravity of their own. As the conflict raged on far beyond the rational calculations of most planners who expected it to last only a few weeks or months, pacifist idealists and some in power realized how much they had underestimated the synergistic power produced by the fusion of nationalism and high technology. The only clear forms visible amid that wild turbulence were the alliances shaped by tangled and fear-driven diplomacy over the previous generation that had led the Great Powers and many smaller nations as well to expand their armies and battle fleets. Few envisioned how much worse it would get—and how fast. As "men of good will" searched for hope in that darkness, many turned to the old rationalist dream of creating an international framework and shaving sovereignty to blunt national rivalries. That vision had been rearticulated in 1910 by the dynamic and quirky Theodore Roosevelt (TR) when he proposed an Anglo-American international police force in his Nobel Prize acceptance speech. His call was not as contradictory to the spirit of that award as it might seem at first glance. Although Alfred Nobel, the prize's creator, had stipulated it should be presented for furthering "fraternity among nations," abolishing or reducing "standing armies," or "the holding and promotion of peace congresses," he doubted the peace movement's ability to influence practical affairs. In the end, he expected the technical evolution of weaponry would terrify mankind into accepting collective security as the mainstay of international order.[13] Roosevelt was, therefore, singing more in tune with his benefactor than he seemed to be at the time.

While TR's proposal made a considerable splash in the media, it was essentially a recasting of Sully's and St. Pierre's visions blended with the ideas of contemporaries and it had no substantial effect. The U.S. Congress passed a joint resolution in late June of that year calling upon President William Howard Taft to form a commission to explore ways to increase arms control and amalgamate the world's fast growing navies to impose peace on the world. Taft, then estranged from his old mentor, did not comply but told Congress in December that he was awaiting responses from other major powers to his queries on those subjects.[14] A veteran lawyer and administrator, he knew how to stifle initiative. Nor did it matter in the larger scheme of things that Roosevelt's speech stimulated a good deal of public discussion and led essayists to draw (rather too easily) analogies between armed forces and civilian police forces. While both Taft and Roosevelt probably realized the latter's suggestions would have little effect in the "real world" of international affairs, TR's neurotic verve gave freshness and bite to what was actually very old wine. In Europe, Roosevelt's proposal was elegantly recast by the prolific pacifist writer von Vollenhoven, who offered a more intricate scheme for protecting small nations against rough handling by major states in future wars,[15] and by Casimir Maciejewski, who urged using international forces to enforce peace.[16] At the same time TR's close friend Rudyard Kipling presented a detailed fantasy of an international air force—IAF—bent on keeping global peace in his short stories "As Easy as A.B.C." and "With the Night Mail." Both lyrically portrayed the use of airships by a benign but coercive international "Aerial Board of Control" to keep the peace through rapid response and intimidation.[17]

Half a decade after Roosevelt's speech, then, his vision gained fresh momentum as World War I raged on far beyond the scale and duration expected by "experts" and led in stages from the League to Enforce Peace (LTEP) and counterpart groups in other countries to the Covenant of the League of Nations, the keystone of the Treaty of Versailles signed after World War I in May 1919. As dozens of designs for systems of international order appeared during and after World War I in succession to a multitude that been devised of the previous half millenium,[18] many observers saw the basic structure of international order itself—what J. M. Spaight called "a world of independent sovereign states"—as the principal nemesis of such projects.[19] As Woodrow Wilson emerged as the great peacemaker, in 1918–1919, the long and intricate lineage of peace enforcement schemes went out of focus. By the end of World War I, let alone during the rest of the twentieth century, few would remember that the first steps along the path toward the League of Nations were taken in early 1915, in an urgent search for a mechanism to preserve peace on both sides of the Atlantic [20] amid unanticipated magnitudes of destruction and misery. The concept of the LTEP was devised, not by Wilson, but by Jane Addams, who proposed the "substitution of an international police for rival armies and navies" to the Woman's Peace Party in January 1915.[21] The LTEP organization was created soon afterward in Philadelphia,[22] and a month later, Addams' resolution was supported by most attendees at the National Peace Conference in Chicago. The concept gained momentum rapidly and was immediately endorsed by thirty-two organizations and three-hundred "distinguished citizens,"[23] including Harvard president Charles Eliot, who called for: "An international police force, subject to the orders of a permanent arbitral court of justice at the Hague" on the grounds that "adequate force must lie behind an international supreme court . . . It may be feared that the court will not command in practice the confidence of civilized mankind."[24]

After a series of Century Club conferences from January to April polished and honed the design of an international court whose coercive power would be exercised by military forces, the soaring popularity of the LTEP led ex-President Taft, after being persuaded by A. Lawrence Lowell to overcome his reservations, to leave the Supreme Court and head the League.[25] The tide of idealistic enthusiasm that crested at Paris had been rising since the Hague Conference of 1915. That well-publicized gathering, whose delegates came from or were sanctioned by the major warring powers, and by the United States and neutral nations as well, had no effect on the battles raging in Europe, Asia Minor, and Africa. Nor was it expected to. Its focus lay beyond the carnage at hand, on the postwar world order, and on drafting plans for a mechanism to stifle or stamp out future wars. While the use of the term "police" in such designs clearly referred to military and naval weaponry, it assuaged the concerns of some pacifists and such prominent Americans as Charles Eliot, who viewed "military organization" as "inconsistent in the highest degree with American ideals of individual liberty and social progress."[26] More important the Hague Conference gave legitimacy and momentum to the LTEP and its counterpart organization in Britain, the League of Nations, and set the tone for such following efforts as the resolution of the U.S. Congress in August 1915, which called on President Wilson to convene an international conference on peaceful settlement and arms control. Like Taft in 1910, he side-stepped that initiative, but by the time the United States entered World War I in April 1917, the idea of a vigorously interventionist league had gained enough momentum that it could not be ignored or finessed, at least not for the moment.

By that point, models of peace enforcement were appearing at all points of the political compass. In Britain, for example, in 1916, the Fabian Society endorsed Leonard Woolf's carefully crafted blueprint for the nations to make "common cause" and in case of aggression "to use naval and military force . . . to enforce the orders of the international high court, by any warlike operations that for the purpose be deemed necessary."[27] In the United States, during the presidential election campaign of that year, each item of what later became the Fourteen Points in early 1918 was addressed in public debates as the president responded to the demands of such groups as the Emergency Peace Federation and the Women's Peace Party.[28] On another quarter, Republican standard-bearer Charles Evans Hughes was promising to lead the nation into the war in Europe, while Teddy Roosevelt shrilly challenged Wilson's will and manhood. Itching to command American troops in France, TR accused the president of cowardice and repeated his earlier proposals for creating an international court whose writ would be enforced by a police force along the lines of an "international *posse comitatus*"—a "World League for Peace."[29] But while he publicly condemned Wilson for not joining other nations in support of the Hague Treaty that protected Belgian neutrality, and for not confronting Germany,[30] Roosevelt privately described G. Lowes Dickinson's 1914 IPF proposal as premature, naive, and founded on ignorance of power processes.[31]

The quickening tide of public enthusiasm for the League to Enforce Peace also brought Taft and Roosevelt back into step. When TR called for a buildup of the nation's defenses, in keeping with the Republican Progressives' Preparedness Movement, Taft concurred, declaring that "nothing but force can cure the brutality and ruthlessness of force."[32] Wilson picked up the cadence while campaigning and co-opted the preparedness theme with a major naval buildup, a mobilization of the National Guard and deployment of most of it and the Regular Army along the Mexican border, and a National Defense Act that was obviously aimed at preparing the nation for a major mobilization. After the United States entered World War I in April 1917, the theme of "a military 'World Police' " became a recurrent one in American political discourse,[33] along with H. G. Wells' "new world order" rubric that would reverberate throughout the twentieth century.[34] As Wilson feared, America's going to war triggered mass hysteria and he himself was not immune to it. In April 1918, speaking on the anniversary of America's declaration of war, he proclaimed, "one response is possible from us: force, force to utmost, force without stint or limit, the righteous and triumphant force which shall make right the law of the world, and cast selfish dominion down in the dust."[35]

There was no shortage of force at hand. By the time the American Expeditionary Force began to form up in France in early 1918, air power had evolved from a gossamer novelty in 1914 to a mailed fist. The Germans had been the first to drop bombs on enemy towns and cities from dirigibles, and later, from heavy bombers, but other major combatants had followed suit with varying degrees of alacrity. Battered by German air raids in early 1917, Britain urged the creation of a major Allied strategic bombing force, but France resisted, concerned that attacks on Germany would be launched mainly from French airfields due to the range limits of Allied bombers at that time. That, they reasoned, might generate reprisal and counterreprisal that would soar out of control, mainly at France's expense. Later in the war, as prospects of a German riposte faded, the French relented, and an Independent Air Force was formed, a multinational bombing force based in eastern France during the last few months of the First World War.[36] Commanded by General Sir Hugh Trenchard, former head of the British army's Royal Flying Corps, this force raided into southeastern Germany, although with less spectacular results than those who cited the Independent Air Force as a precedent—or that he and other IAF enthusiasts claimed later. Critics, including some airmen, saw the Independent Air Force as far more image than substance, and while Trenchard inflated its achievements, he had also been skeptical at first, and took command of it to avoid being put on the shelf. After the war, despite growing optimism in some circles about prospects for imposing peace and international order with air power, there was no consensus about the utility of a multinational bombing force in a "police" peace maintenance role. British military correspondent Colonel Charles Repington, for example, described military aviation as "an auxiliary arm" formed "at the expense of the working infantry," arguing that there was "no evidence . . . that the Independent Air Force shortened the war by an hour."[37] On the other hand, a Royal Air Force (RAF) postmortem asserted that the Independent Air Force's effect on German war production was "known to be great" and its moral impact "enormous" both on civilians and troops at the front.[38]

The bare statistics of the performance were certainly not impressive. The nine squadrons of Trenchard's Independent Air Force dropped about 300 tons of bombs in six months, at a cost of 328 casualties and 352 planes lost, most in ground collisions. There is no way to know if that 1:1 ratio would have held up if the First World War had lasted into the early 1920s, which many expected that it might until the final days of the war. When the armistice was signed in November 1918, Allied planners were preparing to create a much larger Inter-Allied Air Force of some thirty squadrons of British, French, American, and Italian planes and crews. The Independent Air Force's marginal role in the war was obscured after the war when Trenchard inflated its record while serving as Chief of the Air Staff for most of the 1920s as part of his bureaucratic battling with the British Army and Royal Navy who were calling for the RAF's disbandment. Trenchard claimed that a revived Independent Air Force would grow "larger and larger" to "become more and more the predominating factor in all types of warfare,"[39] a vision he passed to the most ardent of all U.S. air power enthusiasts, General William Mitchell, when he visited the IAF's base complex in 1917.

As the growing enthusiasm for a powerful international organization was fed by the grisly spectacle of the seemingly endless war, that in turn led to calls for limiting national sovereignty, like Lowell's suggestion that "the use of force should be automatic instead of subject to the decision of a meeting of the powers."[40] All but 3 of 152 newspaper editorials on the subject in the United States favored using force to preserve international order,[41] and by the spring of 1918, LTEP chapters had been formed in forty-five American states. Beyond the Roosevelt, Addams, and Taft initiatives, a literal host of architects added details and offered suggestions along the way for what ultimately became the Covenant of the League of Nations in 1919. Beyond the deeper and thinner roots of the idea of peace-through-force laid down by Wells and Kipling, Dante, Lord Byron, and Lord Tennyson[42] were the more immediate efforts of William Ladd, Theodore Salisbury Woolsey, co-founder of the American Society for International Law,[43] Elihu Root, Oscar Straus, and Hamilton Holt, who co-edited the *Independent* with Lowell.[44] Others urging such reforms were classicist G. Lowes Dickinson,[45] World Federalism enthusiast Frank Crane,[46] and the redoubtable peace crusader Lord Bryce, former ambassador to the United States and head of the League of Nations Society in Britain during World War I.[47]

Despite the mounting enthusiasm, pacifists remained divided on the issue. Some opposed a collective security system based on military power, like Norman Angell, who rejected using armed force in the service of any cause, no matter how noble it seemed. The relentlessly logical Bertrand Russell favored passive resistance, while others assailed what they saw as the IPF concept's vigilantism, preferring to ostracize and boycott miscreant nations rather than using military power, even the threat of it, to coerce them onto a proper path. Some critics pointed to the old quandary of determining who would have the authority to wield force and condemned LTEP advocates drawing analogies between civil police and armed forces. An especially strained comparison was the suggestion that using international sanctions would be akin to using informal social pressure to bring business practices into line with morality. That was a curious reference considering the American Progressives' struggle in the late nineteenth and early twentieth centuries with monopolies and vendors' sharp practices, including journalistic "muckraking," conservationism, industrial safety laws, antitrust prosecution, and the Poor Food and Drug Act.[48] The crusade for a new world order also enfolded such unlikely members as the arch-skeptic Thorstein Veblen,[49] and various "hard-headed" pacifists.[50] Theodore Roosevelt and elements on the British Left and Center, and especially members of the Union of Democratic Control, mirrored each other in their supporting collective security philosophically while opposing their respective governments' war policies and goals.

Lord Grey, the British foreign minister, while visiting the United States for surgery in 1915, added a semiofficial tone to the rising popular enthusiasm for the LTEP. While his primary goal was gaining President Wilson's assurance that America would support the Allies—"by force if necessary"—after the war when they pressed Germany to return conquered portions of France and Belgium, Grey proposed the creation of "a League of Nations pledged to resist and punish any war begun without previous submission of the cause to international investigation and judgment."[51] That suggestion dovetailed tightly with the LTEP platform and the views of similar groups forming in Europe on both sides of the battle lines,[52] but not with those of the passengers aboard Henry Ford's *Peace Ship*, which sailed for Europe later in the year, pacifists of purer alloy who, like the automobile magnate, opposed using force to reach a higher goal.[53] Other opponents of the war like Norman Angell in Britain, and George W. Kirchwey in the United States, also held that no "good thing is ever accomplished by violence,"[54] while some skeptics anticipated that the inertia of power politics and imperialism would limit the effectiveness of arbitration to resolve minor technical legalities as it had done earlier, and foresaw that an international organization would have no substantial leverage in cases involving the Great Powers' vital interests. The radical Left was also divided. Although many favored pacifism in principle, some supported using force in the form of revolution, or to topple imperialism, like those militant socialists throughout Europe who wanted to form populist armies to enforce the peace, along the lines of Jean Juares' *Nation in Arms*.[55] But they were not so keen on using military forces to maintain international order, fearing that a global system dominated by the big powers would maintain the imperialist and capitalist status quo.

The quandaries on the Left highlighted another dilemma in trying to form an actual League to Enforce Peace—the inequality of nations, not only in terms of differences in population, resources, and wealth, but in respect to imperial holdings. With much of the globe directly ruled by industrialized nations or under foreign economic suasion of the kind that President Taft referred to, however ingenuously, in his phrase "dollar diplomacy," many felt that the basic foundations of imperialism that had been laid down over four centuries earlier, and which were still in place when World War I began, had to a great extent generated the conflict. The United States, Brazil, Argentina, and Russia had extended their contiguous domains, while various smaller nations joined in the "great game" that culminated in the late nineteenth century in a final feeding frenzy in Africa and Asia. Germany, Japan, Italy, and the United States came last to the table to glean small portions left by the major holders of overseas domains—Britain, France, Holland, Spain, and Portugal. The glittering facade over those crumbling mercantilist networks concealed their creeping decay so well that even the sharpest critics of imperialism did not immediately sense how deeply it was being wounded by the First World War. As the conflict dragged on, however, that trauma became more and more apparent, and some shrewd observers, including subject peoples, read the deeper implications of the mother countries' massive armies floundering and bleeding on European battlefields, as well as in Asia Minor and in Africa. Although oppression and wartime censorship in many nations stifled criticism and dissent, the Great War's dragging on led to mutinies and revolts in republics, autocracies, dominions, and colonies alike. The spirit of the times was reflected in the call for self-determination in the Fourteen Points—the list of items for discussion at peace negotiations offered by President Wilson to the Central Powers in January 1918. At the Paris Peace Conference in early 1919, however, a host of such hopes and expectations were confounded. The item of self-determination, and half of those on Wilson's list, would be traded off in exchange for the president's primary goal, the Covenant of the League of Nations. Beyond the Fourteen Points, many expected the

League would include a powerful international police force, but that, too, fell by the wayside in the drafting of the Treaty of Versailles.

NOTES

1. Maximilian A. Muegge, *The Parliament of Man* (London: C. W. Daniel, 1916), p. 92; he also proposed an "international Police Fleet," p. 161.
2. John Hemleben Sylvester, *Plans for World Peace Through Six Centuries* (Chicago: University of Chicago Press, 1943), pp. 1–3.
3. Ibid., pp. 4, 13–16.
4. Quoted in Stefan T. Possony, *Strategic Air Power: The Pattern of Dynamic Security* (Washington, D.C.: Infantry Journal Press, 1949), p. 248.
5. For a concise survey, see Elizabeth V. Souleyman, *The Vision of World Peace in Seventeenth and Eighteenth Century France* (New York: G. P. Putnam's, 1965), p. 91, and M. C. Jacob, ed., *Peace Projects of the Eighteenth Century* (New York: Garland, 1974).
6. E.g., Gondon's *Du Droit Public et du droit des Gens, ou, Principes d'Association Civile et Politique: Suivis d'un Project de Paix Generale et Perpetuelle* (1807); Phiseldek's European League of Nations; the proposal for a European constitution of Friedrich von Schmidt in 1832; and Juan Francisco Sinero in 1842. F. R. Marchand in *Nouveau Projet de Traite de Paix Perpetuelle* suggested Russian, British, Austrian, French, and Prussian contingents under a weighted formula.
7. Quoted in Benjamin F. Trueblood, *The Development of the Peace Idea and Other Essays* (Boston: Plimpton Press, 1932), p. 181.
8. F. H. Hinsley, *Power and the Pursuit of Peace: Theory and Practice in the History of Relations Between States* (Cambridge: Cambridge University Press, 1963), pp. 121–136; also see J. Lorimer, *The Institutes of the Law of Nations* (Edinburgh: William Blackwood, 1883), pp. 448–449.
9. Raoul de la Gresseria, *Des moyens pratique pour parvenir a la suppression de la paix armée et de la guerre* (Paris, 1894); also see Merze Tate, *The Disarmament Illusion: The Movement for a Limitation of Armaments to 1907* (New York: Macmillan, 1942), pp. 119–124.
10. Benjamin F. Trueblood, *The Federation of the World* (Boston: Houghton Mifflin, 1899), pp. 38, 144, 148.
11. Quoted in Charles de Benedetti, *Origins of the Modern American Peace Movement, 1915–1929* (Millwood, N.Y.: KTO Press, 1978), p. 247.
12. For an overview of those times, see Trueblood, *Development of the Peace,* p. 148, and 171–177.
13. Sandi E. Cooper, "Alfred Bernard Nobel," in Harold Josephson, et al., eds., *Biographical Dictionary of Modern Peace Leaders* (Westport, Conn.: Greenwood Press, 1985), pp. 701–702.
14. *Carnegie Endowment for International Peace Yearbook* (Washington, D.C.: Carnegie Endowment for International Peace, 1945), p. 68.
15. See Sandi E. Cooper, *Patriotic Pacifism: Waging War on War in Europe 1815–1940* (New York: Oxford University Press, 1991), pp. 114–116.
16. Casimir Maciejewski, *La Guerre: ses Causes et les Moyens de la Prevenir* (Paris: M. Giard and E. Briere, 1912), p. 77.

17. Rudyard Kipling, "With the Night Mail" [orig. publ. 1905] and "As Easy as ABC" [orig. publ. 1922], both in *A Kipling Pageant* (New York: Literary Guild, 1935), pp. 728–750, and 752–781, respectively; also see "Kipling's Dreams Are Today's Blueprints," *U.S. Air Services*, 28:2 (February 1943): 11, 50.

18. E.g., the proposals of Pierre du Bois, King George (Poděbrad) of Bohemia, William of Ciervia and John Sylvagius, Erasmus, Pope Leo X, Francois de la Noue, Emeric Crucée, Henri IV of France, Cardinal Wolsey, Hugo Grotius, Thomas Hobbes, Bynkershoek, Rachel, William Penn, John Bellers, Abbé de St. Pierre, Cardinal Alberoni, Toze, von Loen, Saintard, Auge Goudar, Johann Franz von Palthen, Jean-Jacques Rousseau, Montesquieu, de la Harpe and Gaillard, von Lilienfeld, Karl Gottlieb Guenther, Bentham, Schindler, Palier de Saint-Germain, Schlettwein, Wolff, Vattel, Kant, Noble, Bertha von Suttner, Vico, Pufendorf, Gondon, von Schmidt, Phiseldek, Sinero, Marchand, Cobden, Alexander II, J. B. Andre Godin, Cornelius von Vollenhoven, Hendrik Dunlop, and Johann Galtung, and such schemes as the 1745 Project and 1787 Peace Projects, the "Holy Alliance" / Concert of Europe / "Congress System," the League to Enforce Peace, the League of Nations, the United Nations, and various regional Cold War aggregations. This list is derived from Muegge, *The Parliament of Man*, Jacob Ter Meulen, *Der Gedanke der Internationalen Organisation in seiner Entwicklung* (The Hague: Martinus Nijhoff, 1917 [1968]), pp. ix–x; Elizabeth York, *Leagues of Nations, Ancient, Mediaeval, and Modern* (London: Swarthmore Press, 1919); Lord Phillimore, *Schemes for Maintaining General Peace* (London: H. M. Stationery Office, 1920); S. C. Vestal, *The Maintenance of Peace or the Foundation of Domestic and International Peace as Deduced from the Study of the History of Nations* (New York: G. P. Putnam's, 1923); Merle Curti, *Peace or War: The American Struggle, 1636–1936* (New York: W. W. Norton, 1936); Stefan T. Possony, "Peace Enforcement," *Yale Law Journal* 55 (August 1946): 910–949; Hinsley, *Power and the Pursuit of Peace*; Leonard William Doob, *The Pursuit of Peace* (Westport, Conn.: Greenwood, 1981); Albert Lepawsky, Edward H. Buehrig, and Harold D. Lasswell, eds., *The Search for World Order* (New York: Appleton-Century Crofts, 1971); Seyom Brown, "Efforts to Purge War from the World" and "Efforts to Reduce the Role of War in the International System," *The Causes and Prevention of War in the International System* (New York: St. Martin's Press, 1987), pp. 115–172; and Charles Chatfield and Ruzanna Ilukhina, *Peace/Mir: An Anthology of Historic Alternatives to War* (Syracuse: Syracuse University Press, 1994).

19. J. M. Spaight, *An International Air Force* (London: Gale & Polden, 1932), p. 17; for various perspectives, also see Edward Krehbiel, *Nationalism, War, and Society* (New York: Macmillan, 1916), and Martin Wight, "The Balance of Power and International Order," in Alan James, ed., *Bases of International Order* (London: Oxford University Press, 1973), pp. 85–115.

20. E.g., the "Platform of the League to Enforce Peace," *Survey* 34:28 (June 16, 1915): 291.

21. Krehbiel, *Nationalism, War and Society.*

22. De Benedetti, *Origins of the Modern American Peace Movement, 1915–1929*, pp. 4–6, 2–26, 53–54; Randolph S. Bourne, *Towards an Enduring Peace* (New York: Charles Scribner's Sons, 1916), pp. 264–265; and C. Roland Marchand, *The Peace Movement and Social Reform* (Princeton: Princeton University Press, 1972), pp. 181–182.

23. Denna Frank Fleming, *The United States and World Organization 1520–1933* (New York: Columbia University Press, 1938), pp. 11–13.

24. Charles W. Eliot, *The Road Toward Peace* (Boston: Houghton Mifflin, 1915), p. 25; also see Eliot's "How Can America Best Contribute Toward Constructive and Durable Peace?" *Annals of the American Academy of Political and Social Sciences* 61 (September 1915): 244.

25. Howard F. Pringle, *The Life and Times of William Howard Taft* (New York: Farrar and Rinehart, 1939), 2:927–928.

26. Eliot, *The Road Toward Peace*, pp. 14, 67.

27. Leonard S. Woolf, *International Government: Two Reports* (New York: Brentano's, 1916), p. 410.

28. For an overview see Curti, *Peace or War: The American Struggle 1636–1936*, pp. 240–248.

29. Theodore Roosevelt, *America and the World War* [orig. publ. 1916] and *God Take Your Own Part* [orig. publ. 1915] (New York: Scribner's, 1926), pp. 53, 71.

30. Ibid., pp. 75–77.

31. Elting Morison, ed., *The Letters of Theodore Roosevelt* (Cambridge: Harvard University Press, 1954), 8:852.

32. William Howard Taft, "Victory With Power," in *Win the War for Permanent Peace* (New York: League to Enforce Peace, 1918), p. 17.

33. E.g., Christen Collin, *The War Against War and the Enforcement of Peace* (London: Macmillan, 1917).

34. Richard Koebert and Helmut Den Schmidt, *Imperialism: The Story of a Political Word, 1840–1960* (Cambridge: Cambridge University Press, 1964), p. 329.

35. *President Wilson's Great Speeches and Other History Making Documents* (Chicago: Stanton and Van Vlies, 1919), p. 354.

36. For a basic account of the Independent Air Force, see Chaz Bowyer, *History of the RAF* (London: Basic Books, 1977), p. 40; one sympathetic to Trenchard is Andrew Boyle, *Trenchard: Man of Vision* (New York: Norton, 1962), pp. 255–318; highly critical is Benjamin D. Foulois, "Air Service Lessons Learned During the Present War," in Maurer Maurer, ed., *The U.S. Air Service in World War I* (Washington, D.C.: Office of Air Force History, 1979), 4:26–27; and contemporaneous organizational perspective is PRO AIR 1/461, "History of the Organization for Bombing Germany . . . ," n.d., c. October 1918.

37. Boyle, *Trenchard*, p. 352.

38. PRO 1/457, "Bombing of Germany by British Aircraft—Oct. 1, 1917–Nov. 11, 1918," n.d.

39. Boyle, *Trenchard*, p. 351.

40. Ruhl J. Bartlett, *The League to Enforce Peace* (Chapel Hill: University of North Carolina Press, 1944), p. 36.

41. Sylvester, *Plans for World Peace Through Six Centuries*, p. 155.

42. For various genealogies, see David Starr Jordan, *Ways to Lasting Peace* (Indianapolis: Bobbs-Merrill, 1916); Charles Chatfield, *For Peace and Justice: Pacifism in America 1914–1941* (Knoxville: University of Tennessee Press, 1971), pp. 10–11; and George Schwarzenberger, *Power Politics: An Introduction to the Study of International Relations and Post-war Planning* (London: Jonathan Cape, 1941), p. 351.

43. Theodore Salisbury Woolsey, "War Thoughts," *North American Review* 201: 1 (January 1915): 52–53.

44. See Hamilton Holt, "The Constitution of a League," *Independent*, September 28, 1914, pp. 160–163, and the elaboration of his ideas in *Selected Articles on World Peace* (White Plains, N.Y.: H. H. Wilson, 1916), pp. 207–213.

45. G. Lowes Dickinson, "How Can America Best Contribute to the Maintenance of the World's Peace," *Annals of the American Academy of Political and Social Sciences* 61 (September 1915): 235–268; and Jordan, *Ways to Lasting Peace*, pp. 79–86.

46. Frank Crane, *War and World Government* (New York: John Lane Co., 1915).

47. Curti, *Peace or War: The American Struggle 1636–1936*, p. 238; for a perspective on Bryce's thinking at the time, see Viscount Bryce, et al., *Proposals for the Prevention of Future Wars* (London: George Allen and Unwin, 1917).

48. E.g., see Robert Goldsmith, *A League to Enforce Peace* (New York: Macmillan, 1917), pp. 132–136.

49. Thorstein Veblen, *An Inquiry into the Nature of Peace and the Terms of Its Perpetuation* [orig. publ. 1917] (New York: The Viking Press, 1945), esp. pp. 230–232.

50. E.g., Lucia Anne Mead, "Primer of the American Peace Movement," in *Selected Articles on World Peace* (New York: H. H. Wilson, 1916), p. 113.

51. Henry Van Dyke, *Fighting for Peace* (New York: Scribner's, 1917), pp. 241–242.

52. E.g., see *New Republic* editorials of May 20 and June 26, 1915, and Woolf, *International Government*, esp. 371–379, and 409; also see J. A. Hobson, *Toward International Government* (New York: Macmillan, 1915), p. 182.

53. E.g., John Haynes Holmes' assailing of the "Gospel of Force" in *New Wars for Old* (New York: Dodd and Mead, 1916).

54. George W. Kirchwey, "How America May Contribute to the Permanent Peace of the World," *Annals of the American Academy of Political and Social Sciences* 61 (September 1915): 231.

55. A clinical perspective is Marvin Swartz, *The Union of Democratic Control in British Politics During the First World War* (Oxford: Clarendon Press, 1971); an excoriation of Angell and the Union of Democratic Control (UDC) is G. G. Coulton, *The Main Illusions of Pacifism: A Criticism of Mr. Norman Angell and the Union of Democratic Control* (Cambridge: Bowes and Bowes, 1916).

II

Bloom and Wilt: The International Police Force Concept and the League of Nations

Despite rapid technical advances in the development and construction of dirigibles and airplanes after 1900, on the eve of the First World War, most military professionals saw aircraft as little more than gadgets. From that point on, as debates between enthusiasts and skeptics regarding air power's utility raged in the armed services of many nations throughout the twentieth century, a myriad of plans for harnessing air power to various purposes appeared. As a wave of aviation mania swept the world—what Joseph Corn deemed the "winged gospel"—those who proposed creating an international air force, like earlier proponents of using force to keep the peace, found themselves mired in a morass of paradoxes. As noted earlier, despite a host of attempts to draw analogies between armed forces and police forces, many still saw using force for any purpose as wrong, even if under a legitimate international authority. Some of those saw force as evil in itself, while others opposed allowing an international authority to wield weapons which might later be outlawed by arms control agreements, or which might be used to coerce individual nations. Considering that tangle of views and motives, it is not surprising that all diplomatic attempts to ban or limit the use of air bombing during the two decades before World War I failed.

Before the First World War, some analysts had explored the potential of air power, examining such diverse aspects as the legality of air attacks on defense installations located in populated areas and air defense tactics.[1] The first dropping of bombs from airplanes in war—by the Italians against Turkish forces in Tripoli in 1912—stirred up some concern, but the combined air strength of the major combatant nations on the eve of World War I in 1914 was fewer than a thousand planes. In the tactical realm, military and naval professionals saw aircraft mainly as a means of reconnaissance.[2] While the fledgling air forces played that role effectively in the opening campaigns of the Great War, it soon became apparent that aircraft had much greater potential. Bomber, fighter, reconnaissance, and transport types soon appeared in large numbers. Millions throughout Europe and much of the world saw flickering newsreel images of the rapidly evolving air war. Although those scenes were usually posed or faked, like most war photographs and magazine and newspaper illustrations, they meshed with air power enthusiasts' grandiose hopes of winning military victories quickly and cheaply. Although the mounting losses of airmen, and the deaths of perhaps 7,000 civilians in bombing raids during 1914–1918 seemed horrifying at the time, that scale of casualties was very small compared to what was to follow.[3]

When the Great War ended, Europe and much of the world were in a state of turmoil and shock, although the rending of the old order and the eruption of mutinies and revolutions led many to fear or hope for new forms and patterns of politics, social values, and even the reordering of the world on a global scale. But despite the great upwelling of optimism that attended the opening of the Peace Conference in Paris in early 1919, imperial subjects' aspirations and entreaties had little effect on the shaping of the Versailles treaty. Many colonies received new sets of chains under the mandate system; their plight was overlooked or left out-of-focus by designers of various plans for "policing" the world. In one detailed proposal, for example, "aerial police" contingents under a "constabulary office" would maintain the peace, and serve as the "beginning of international police administration" intended to open the way for similar activities on "land and water," including "an international Police Fleet."[4] Such schemes, which critics dismissed as new ways to enforce the status quo, also came to naught. So did blueprints for universal disarmament calling for all nations to transfer their arms "completely to an international police force," which would serve under an international tribunal "for the better ordering of mankind."[5] The details of just what the application of force might look like were rarely visible in such blueprints beyond general statements.

A great many widely recognized lessons of the Great War soon faded from view as the Paris conference began. Despite the one-sided terms of the armistice, millions still hoped the peace treaty would encompass the Fourteen Points, even though popular support was declining in the United States by the war's end, a problem that Wilson was forced to grapple with on the way to Paris and during the conference. On the eve of the peace deliberations, many observers saw Wilson, as a former professor and not a professional politician or diplomat, as the least tainted by the stains of power-seeking among the "Big Four" Allied heads of state. The United States, most modern, largest, industrially powerful, and least imperialistic of the Great Powers—and most remote and least damaged by war, had come last to the table in the game of power politics, and with a relatively small stack of chips. That led British scholar G. Lowes Dickinson to assert that the United States was unique among the nations involved in shaping the peace, and the "only one likely sincerely to take the view of the peoples instead of the militarists and diplomats."[6] In a similar vein, Wilson claimed the American delegation was the only disinterested party at the peace conference and that its single major goal was the framing of some sort of covenant.[7] That rang a bit hollow, if not disingenuous, since he would effectively defend the Monroe Doctrine and other American agendas at Paris, even though his domestic political flank was open from late 1918 onward. In the congressional elections of that year, Wilson had asked the American electorate to vote for a Democratic Congress to support his position in shaping the peace. The loss of Democratic seats was compounded by the distrust of Republicans in Congress who doubted his judgment in going to the peace conference, a decision that concerned other Americans as well, and in not giving the Grand Old Party (GOP) more than a very small presence in the delegation.

Then and since, Wilson's defenders deemed his attendance as crucial in assuring that the Covenant of the League of Nations and peace treaty were shaped in the proper spirit, since the French, British, and Italians were bent on pursuing their national interests, as they had throughout the war.[8] Whatever American sentiments may have been, Wilson was widely acclaimed in Europe, which raised the hopes of millions there during the few weeks between the armistice and the peace conference that the visionary American president might redeem the vast suffering and slaughter of the war by creating a new world order that would prevent another such conflagration. In contrast to the vast crowds who greeted Wilson's grand processional through France, Britain, and Italy on his way to the conference table, however, a sense of gloom suffused the Paris gathering from the outset, reflecting war-weariness and bitterness.[9] The Germans were excluded from critical deliberations, and much of the business was conducted by the major Allied powers in secret, contrary to the Fourteen Point item calling for "open agreements, openly arrived at." The abandonment of that and other points fed the sense of disillusionment as it soon became obvious that it was to be a victors' peace and that the American president's expectations were being thwarted by a reversion to traditional diplomacy.

The pivot of the discussions at Paris and the Versailles treaty was the Covenant of the League of Nations, initially envisioned as a powerful international structure. Before the conference began, Wilson, many delegates, and millions throughout the world favored a "real central international authority" which would "be bound to take action against any breach of the peace . . . by any means in their power . . . in the last resort by force of arms" including an international force supported by a tax on the nations.[10] Opponents included French generals, certain ethnic minorities and imperial subjects throughout the world, and many Americans beyond the Republican "irreconcilables" in the Senate. Herbert Spencer, for instance, declared that the use of military methods against war was a move toward "rebarbarization,"[11] while others echoed Frederick the Great's disdain in describing the idea of an armed league enforcing peace as "a fatuous dream."[12] There was also a good deal of irony in Wilson championing the use of armed force to any purpose, given his aversion to military matters and distance from initial attempts to create an international collective security system. To various supporters and detractors, his co-opting the League concept and sealing many of its originators off from the peace process demonstrated willfulness, political skill, or his sense of personal destiny. In any event, his oft expressed disdain for military affairs did not ring true in respect to his presidential actions. Like a fastidious lord of the manor, he did not visit the kennels, but ordered hounds loosed when he deemed it necessary, dispatching several expeditionary forces in the western hemisphere, and signing off on the biggest peacetime naval construction program in U.S. history up to that time.[13] In the 1916 election campaign, he carried a flag in a Preparedness movement parade, endorsed the League to Enforce Peace's platform, and signed the National Defense Act that laid the foundations for mass mobilization. All that stood in contrast with the neutralist stance embodied in his first secretary of state William Jennings Bryan. In the early months of World War I, Bryan, a professed pacifist despite a brief stint in uniform during the war with Spain,[14] arranged unilateral treaties with twenty-one European nations providing for the arbitration of any mutual dispute. While that intricate diplomatic frieze had no practical effect in the long run, it threw a revealing light on the U.S. Senate's refusal to ratify the Covenant of the League of Nations in 1920. Bryan's treaties, approved by the Senate, authorized five-member arbitration panels, including one U.S. member, weakening American national interest far more than the Covenant.[15] True to his principles, Bryan resigned in the late spring of 1915 over Wilson's condemning Germany for the sinking of the *Lusitania*, then engaged Taft in a verbal battle that raged until America went to war, during which Bryan assailed the League to Enforce Peace's proposals for an international police force.

Wilson stayed clear of that duel while taking close control of foreign affairs in close counsel with his aide Colonel Edward House. Presenting himself as a disinterested arbiter of the peace, a role he still strove to play even after becoming a wartime commander-in-chief in April 1917, he brought the United States into the war not as an Allied, but as an "Associated Power." Struggling to remain above the battle in early 1918, he offered the Fourteen Points to the Central Powers as the basis for the armistice. Buoyed up by sympathetic media hyperbole and cheering crowds in Europe, Wilson, like many others of an optimistic bent, discounted the congressional elections of the 1918's going against the Democrats, since they had been held just before the war ended suddenly. In any case, the president soon began veering, first by deciding to break precedent and go to Europe, then by including only one Republican in his entourage when he left for Europe at the end of the year, and then by abandoning the IPF concept in midvoyage. Halfway across the Atlantic, he asked his shipmates, the Executive Council of the Body of Delegates, to prepare recommendations on how "armed forces to be used against the convention-breaking state" should be configured, including the contingency of using "any force that may be necessary" to close frontiers.[16] For whatever reason, Wilson's declining health, the isolation of the sea voyage, or his advisors' dim view of a "Grand Police Alliance,"[17] the president now fell in step with others among the elites of the Allied camp who were urging moderation in the face of popular enthusiasm for an IPF.[18] A month earlier, for example, just as the war ended, the eminent British peace advocate Lord Robert Cecil pointed to the League to Enforce Peace's original platform when he asserted that the signatories of a peace pact that included such a structure would not be establishing an international police force, nor promising to punish any nation which resorted to war. Only those who undertook hostilities without first submitting the precipitating dispute to a conference were to be coerced. In a similar vein, Field Marshal Jan Christiaan Smuts of South Africa, in a pamphlet entitled *The League of Nations: A Practical Suggestion*, opposed establishing an international military mechanism to enforce any agreement that might be reached at the peace conference. He saw a league of nations—the popular term in Britain at the time—as an organization "based on a common will to maintain the peace," not "on differences" but on the sense of "a community of power, with no immunity for anyone from its obligations and responsibilities."

Those adjudging Smuts' and Wilson's broad and vague rhetoric as naive, for example, the call for "an inner transformation . . . an inner change" in international affairs that would transcend specific mechanical arrangements, were soon proven right.[19] Although Wilson was raised in the American South after the Civil War, he gave little weight to parallels between that struggle and World War I—each lasted far longer than expected, cost far more in lives and treasure, amplified hatreds and resentments, and ended with a victors' peace that left major imbalances unresolved, and which engendered further pathologies. Indeed, the French, most bent on vengeance but also more realistic about security issues, played the role of the American Radical Republicans during the Civil War and Reconstruction. As the delegates of many small nations and lobbyists orbited the "Big Four"—Prime Ministers Lloyd George of Great Britain, Clemenceau of France, and Orlando of Italy and President Wilson—seeking crumbs from their table, the peace negotiations drifted far from the discussions among a community of equal nations implied in Wilson's wartime rhetoric and the Fourteen Points. In the spirit of Mao Ze Dong's adage that "power grows out the barrel of a gun," the French kept substantial forces on active service, while the British and Americans speedily disbanded theirs, putting the peace conference literally in the shadow of guns.[20] Wilson soon found that the wild ovations that marked his meandering path to the Paris meeting, which many saw as presaging the end of traditional diplomacy, provided no solid constituency as he grappled with powerful national interests. Questions like legitimacy, sovereignty, the war-guilt question, imperialism, disarmament, and meting out rewards and punishments all took on different hues in the somber setting of Paris than they had in the glare of the war.

Wilson, already the main focus of praise and condemnation at home and in Paris as proceedings began formally on January 18, 1919, found himself beset by throngs of supplicants from across the world, including, as a senior British police official noted with alarm, "British Bolsheviks, Catalan separatists, French Majoritarian Socialists, Irish Sinn Fein, [and] Indian anarchists."[21] Their imprecations had little effect, but did feed the president's concern about substituting "international militarism for national militarism" and underscored the old quandary of maintaining the international status quo in the name of preserving order and peace.[22] Despite the United States' disdain for mandates, League of Nations' trusteeships over former German and Turkish colonies awarded to various Allied nations, America's special interests were recognized, most notably in provisions leaving the Monroe Doctrine intact, condemned by a critic as "the most absolute, uncompromising denial of the right of interference by an international police as a means of keeping the domestic peace."[23] As the IPF concept faded from view, entreaties from most of the great empires' subject peoples for self-determination were ignored or rejected, along with Japan's resolution supporting racial equality.[24] Some Latin American nations' objections to various provisions were not translated for the record.[25]

Despite several American successes, a folklore developed that the U.S. delegation's poor preparation contrasted with other major powers teams of technicians and area specialists, and their sophisticated background papers, putting it in the position of American country boys being duped by European city slickers. That meshed with another popular misperception, that Wilson haggled away all of the Fourteen Points to retain the Covenant, allowing France to have her way with Germany and generating the resentment in Weimar Germany that nourished the growth of National Socialism. In fact, Wilson managed to retain half the items on the list and to fend off France's demands for a solid, militarily powerful league of nations, not small achievements considering the French generals' fervor, their large military forces nearby, and the conference's location in Paris. France, having suffered far more war damage and far more casualties than the other western Allies, saw "the problem of the Organization of the Society of Nations" more in terms of "a military problem"[26] and presented detailed proposals on issues like inspection and supervision of the League's military elements, and whether such forces should be standing forces or assembled from contingents maintained by member nations. With support from some other nations, especially Spain, France would continue to urge the creation of a formal League military structure and an international general staff until World War II,[27] but in pressing for a well-founded international military apparatus, they misestimated the longstanding Anglo-American suspicion of Prussian models and French motives. Only the American and British armies' respective bungling in the Spanish-American and Boer Wars had led them to create even weak facsimiles of a Prussian general staff.[28] The Anglo-Americans were also more optimistic than the French about the ability of economic sanctions to bring aggressors to heel, although that was not wholly off the mark, since many then saw the Allied blockade as a major factor in the Allied victory, and it was still being maintained as a coercive tourniquet on Germany during the peace conference, despite many protests around the world.

The draft of the Covenant of the League of Nations submitted to the Plenary Session of the Paris conference in mid-February was mainly the work of an American, David Hunter Miller, and of Britain's Lord Cecil. At that point, Wilson was locked in battle on another front, as growing opposition to the Covenant in Congress led him to return to the United States late that month. The president's tense confrontation with congressional leaders over dinner led thirty-nine Republican senators to sign a "Round Robin" statement in early March insisting that the Covenant be separated from the peace treaty. After Wilson declared that such an attempt to "dissect the Covenant from the treaty" would destroy "the whole vital structure" of the peace, he returned to France, where he came under fire from Marshal Foch. The generalissimo demanded that Germany pay punitive reparations to France and Belgium and called for firmer guarantees to keep the Rhineland out of German hands. His health failing visibly, the president threatened to return home, throwing Foch and his colleagues back in disarray. At that point, the Italian delegation charged in, making territorial claims that led Wilson to appeal to the Italian people over the heads of their leaders. That forced Premier Orlando to withdraw and return later without gaining his demands.

While the Japanese were able to gain control of former German holdings in China, Wilson managed to check their Siberian ambitions. At that point, the president accommodated some pro-League Republicans, most notably Taft, with Covenant amendments excluding the United States from the mandate system, limiting League control of tariffs and immigration, recognizing local arrangements "like the Monroe Doctrine," and providing for American withdrawal on two years' notice.

Beyond those adjustments, the Covenant had been shaped and reshaped in the horse-trading of traditional diplomacy, so that everyone received less than a full loaf. France, for example, got reparations and control of the Rhineland but not the solid structure of an international police force that she sought. The Americans and British saw French proposals for an international general staff and standing League armed forces at Paris as attempts to dominate Europe and render Germany a vassal, discounting the possibility that more might be at stake than Gallic territoriality and vengefulness. When French delegates presented a list of measures, labeled "a" through "m," the Americans and British interpreted the last item—"joint war on the recalcitrant state by all loyal members"[29]—as a way to create a "Ring of Steel" around Germany, too close to power politics and too far from an equitable model collective security.[30] Throughout the discussions, as the French concentrated on the European balance of power, British and American resistance to their entreaties led French delegate Léon Bourgeois, author of *Pour la Société des Nations* (1910) and a dedicated proponent of a well-armed League, to predict that the failure to create a standing army had doomed the League to failure. Even though some Americans agreed with the view that "there must be security before there can be disarmament,"[31] France's steady pounding on the issue irritated their colleagues and led Wilson, in exasperation, to threaten to form an Anglo-American alliance in lieu of the Covenant.[32]

Although the first draft of the Covenant looked much like the League to Enforce Peace's 1915 agenda in requiring that disputes be submitted to an international tribunal and proposing a council of reconciliation, subsequent drafts thinned out those provisions. An early modification, for example, called for members' use of "both economic and military force" if a League member nation went to war without referring the dispute to the council for resolution and for codifying and improving international law to make such procedures easier,[33] but even that was later rejected. As Wilson braked the momentum toward a "League with teeth," his view on using force as a "last resort" contrasted with both his pronouncements in the spring of 1916 that "the peace of society is obtained by force" and that "back of opinion is the ultimate application of force" and his insistence immediately after the war that the "organized major force of mankind" must be so great "that no nation, or probable combination of nations, could face or withstand it."[34] That reversal, of course, reflected political trends and maneuvers in the United States. Even when Wilson abandoned his earlier strong stand by proclaiming that the Covenant was "a Constitution of peace, not . . . a league of war,"[35] his political adversaries still claimed he was trying to connect the United States to a hair-trigger collective security commitment.

There were other political barriers across Wilson's path. A second term president, he could not run again under the informal but very solid two-term limit then in effect, although the Democratic Party would have gained substantial prestige if he brought home the very large slab of diplomatic bacon that ratification of the Covenant would have represented. Worse, the struggle over the Covenant had become a blood political fight. As his opponents dug their trenches deeper, he followed suit and insisting that the American public and Congress support the Covenant without significant qualification. After all, the provisions of that document regarding the League's use of force lay very far from the early stages of the Paris conference, when a standing international force of some five million troops had been part of the design. And while Theodore Roosevelt saw Wilson's vision, including the Fourteen Points, as idealistic rubbish, he and other Republicans retained their enthusiasm for an international police force, along with some German Social Democrats and others in the liberal camp in Germany.[36] During the war, Matthias Erzberger envisioned national contingents on call for expeditionary service like the Allied armies of World War I and the Boxer Expedition of 1900[37]—contrary to German militarist propaganda portrayals of such forces as bullies under the rubric "*Eine Ganze welt gegen uns*!"—a whole world against us.

In the end, those fragments of enthusiasm made no difference. The biggest hurdles to American involvement in the League of Nations proved to be those that were in plain view when Wilson sailed for Europe—the U.S. Constitution's provisions that Congress must declare war and that the Senate must ratify any treaty. Both prevented committing the United States to employ military force under either predefined or uncertain conditions. Both advocates and critics of the Covenant alike realized from the outset that the Senate might reject the peace treaty or pare it down to a thin reed. Wilson's bargaining some of the Fourteen Points against the Covenant did not sway the "Irreconcilables" in the Senate nor other opponents among American elites, the media, and the public-at-large. Press accounts of the Paris conference portrayed the peace negotiations sliding back to traditional backdoor and big stick diplomacy and producing a punitive and unbalanced peace treaty. Might Wilson have brought the Covenant through if he had been as willing to compromise at home as he had been in Paris? His abandonment of the League to Enforce Peace's original position became plainly apparent early in the conference that began to drift from its original moorings began when France's proposal for a strong, credible military response force under League control was voted down on February 13, 1919.[38] Wilson met French objections with assurances of "good faith" and "trust," promising that France and other nations would have plenty of time to react to any German resurgence, at which point "the whole world would be ready to vindicate its liberty."[39] Plainly apposite to the U.S. Constitution, that assurance showed the French his lack of concern for their interests while heightening his opponents' fears in the Senate.

In the course of the labyrinthian peace process, the document initially labeled Protocol 5 evolved into the Covenant of the League presented in a plenary session on April 28.[40] Some original forceful statements remained, including the option of acting against nonmember states in Articles 7 and 8, complex and ambiguous rules for using force in Articles 10, 16, and 17, and a mention of the contingency of "combining their armed forces on its behalf."[41] Article 16, Sec. 2 came closest to the IPF issue by providing that in cases of aggression, the League Council could recommend to members "what effective military, naval or air forces the members of the League shall severally contribute to the armed forces to be used to protect the Covenant of the League." Despite the thinness of the final product, many still hoped it would lay the foundation of an ultimate mighty structure of international order to be erected over time on layered precedents of international law, cooperation and, if need be, force. However, others saw that as unlikely as Frederick the Great had, given the hardihood of sovereignty, or opposed creating an international peace system on violence. But their concerns proved irrelevant, for only the faintest traces of the burning enthusiasm for creating an IPF that rose from 1915 to 1918 remained in the Versailles treaty signed on June 28, 1919.

Wilson made his final transatlantic journey, and on July 10, set the agreement before the Senate, which was now split into three camps on ratifying the Covenant: the protreaty Democrats mustered by Gilbert Hitchcock of Nebraska; "moderate" followers of Foreign Relations Committee chairman Henry Cabot Lodge, Massachusetts Republican; and the "Irreconcilables," the most prominent of whom, William Borah of Idaho, Hiram Johnson of California, and Robert La Follette of Wisconsin, were in a majority on the Foreign Relations Committee. The latter group rejected Wilson's mid-August offer to accept informal reservations that would not be included in the ratification resolution text. After introducing forty-nine modifications to freeze the treaty in the Senate till mid-September, Johnson and Borah toured the country to assail the League, in a railroad jaunt subsidized by magnates Henry Frick and Andrew Mellon. As Wilson tried to counter the Irreconcilables by making his own train trip, his health, visibly fragile since the Paris conference began, failed completely in December. Totally "out of the loop" when the treaty was first formally considered by the Senate, Wilson rallied, and despite bipartisan support for Lodge's moderate ratification resolution, demanded unconditional passage. That led both Democrats and Irreconcilables to vote against ratification, rendering the Covenant hollow, and returning the IPF concept to the realm of visionary speculation, where it would remain for the next fifteen years.

NOTES

1. See J. M. Spaight, *Aircraft in War* (London: Macmillan, 1914), p. 22, and F. W. Lanchester, *Aircraft in Warfare: The Dawn of the Fourth Arm* (New York: D. Appleton, 1916), Sec. 119–1.
2. John Howard Morrow Jr., *Building German Airpower, 1910–1914* (Knoxville: University of Tennessee Press, 1976), pp. 86–87.
3. The largest air raid of World War I involved fifty aircraft, and the Germans' biggest attack on London, only thirty-six. Less than 300 British civilians were killed versus 62,000 in World War II, while Germany lost 746 and over half a million respectively in those conflicts—the latter the result of 16,000 tons of bombs dropped by the Allies 1914–1918, versus 2.7 million tons from 1939 to 1945; Edward Mead Earle, "The Influence of Air Power Upon History," *Yale Review* 35:4 (June 1946): 583.
4. Maximilian A. Muegge, *The Parliament of Man* (London: C. W. Daniel, 1916), pp. 118, 161, 174.
5. John Haynes Holmes, *New Wars for Old* (New York: Mead and Co., 1916), p. 262; also see H. E. Hyde, *The Two Roads: International Government or Militarism* (London: P. S. King, n.d. [1916]), pp. 7–20.
6. G. Lowes Dickinson, "How Can America Best Contribute to the Maintenance of the World's Peace," *Annals of the American Academy of Political and Social Sciences* 61 (September 1915): 235.
7. James T. Shotwell, *At the Paris Peace Conference* (New York: Macmillan, 1937), pp. 75–77.
8. Charles Seymour, *Geography, Justice and Politics at the Paris Peace Conference, 1919* (New York: American Geographic Society, 1951), p. 18.
9. E.g., see Arthur Schlesinger Jr., *The Age of Roosevelt: The Crisis of the Old Order, 1919–1933* (Boston: Houghton Mifflin, 1956), p. 14, and Harold Nicolson, *Peacemaking 1919* (New York: Grosset and Dunlap, 1965).
10. See "The League of Peace Treaty," in Severin Nordentoft, *Practical Pacifism and Its Adversaries* (New York: Frederick A. Stokes, 1918), pp. 179–185.
11. Denna Frank Fleming, *The United States and World Organization, 1920–1933* (New York: Columbia University Press, 1938), pp. 13, 38–40.
12. See W. Morgan Shuster, "Peace and Disarmament," *Century* 89:2 (February 1915): 503–511.
13. David Lloyd George, *Lloyd George: A Diary* (New York: Harper and Row, 1971), p. 175.
14. William H. Taft and William Jennings Bryan, *World Peace* (New York: George H. Doran, 1917), esp. pp. 79, 87, and 173.

15. Livingston Hartley, *Is America Afraid?* (New York: Prentice-Hall, 1937), p. 390.

16. Robert Lansing, *The Peace Negotiations: A Personal Narrative* (Boston: Houghton Mifflin, 1921), p. 287.

17. For some informal perspectives, see A. J. P. Taylor, ed., *Lloyd George: A Diary by Frances Stephenson* (New York: Harper and Row, 1971), p. 172, 277.

18. Alfred Ollivant, *The Next Step: An Essay on the Missing Policeman* (London: George Allen and Unwin, 1919), p. 85.

19. An analysis of the impact of Smuts' pamphlet is F. P. Walters, *A History of the League of Nations* (London: Oxford University Press, 1952), pp. 27–28; also see H.W.V. Tamperley, ed., *A History of the Peace Conference of Paris* (London: Oxford University Press, 1969), 3:51.

20. Marshal Foch ordered that captured German artillery pieces be parked on the streets of Paris, symbolizing his disdain for the League concept.

21. Basil Thomson, assistant commissioner of Metropolitan Police, and head of the CID, quoted in Marvin Swartz, *The Union of Democratic Control in British Politics During the First World War* (Oxford: Clarendon Press, 1971), p. 219.

22. David Hunter Miller, *The Drafting of the Covenant* (New York: G. P. Putnam's, 1928), 1:294, and Sondra Herman, *Eleven Against War: Studies in American Internationalist Thought, 1898–1921* (Stanford: Hoover Institution Press, 1969), p. 208.

23. S. C. Vestal, *The Maintenance of Peace Or the Foundation of Domestic and International Peace as Deduced from the Study of the History of Nations* (New York: G. P. Putnam's, 1923).

24. Oscar Newfang, *United States of the World* (New York: G. P. Putnam's, 1930), p. 198.

25. Charles Mee Jr., *The End of Order: Versailles, 1918* (New York: E. P. Dutton, 1980), p. 200.

26. General [Pseudonym for Victor Marguerritte], *Organisation Militaire de la Societe des Nations* (Paris: Editions Habdomadaire, c. 1919), p. 8.

27. Miller, *Drafting of the Covenant*, 1:11–12, 42, 243–259; for an advocacy of the contingent model as opposed to an IPF see W. G. S. Adams, "The Basic of Constructive Internationalism," *Annals of the American Academy of Political and Social Sciences* 61 (September 1915): 220.

28. E.g., Charles Walston, "League of Dreams or League of Realities? A Supernational Jury and Police Force?" in *The English-Speaking Brotherhood and the League of Nations* (New York: Columbia University Press, 1919), pp. 209–224.

29. See French Document–Annex 2 to the Minutes of the First Meeting, French Ministerial Commission for the League of Nations, pp. 243–244 and 323–324, in Miller, *The Drafting of the Covenant*, vol. 2.

30. Frank G. Williston and Linden A. Mander, "The Problem of Security," in Joseph B. Harrison, Linden A. Mander, and Nathaneal H. Engle, eds., *If Men Want Peace: The Mandates of World Order* (New York: Macmillan, 1946), p. 20.

31. Harold Josephson, *James T. Shotwell and the Rise of Internationalism in America* (Rutherford, N.J.: Fairleigh Dickinson University, 1975), p. 118.

32. Miller, *Drafting of the Covenant*, 1:253.

33. A. L. Lowell, "League to Enforce Peace," *Atlantic Monthly* 116:9 (September 1915): 393–394.

34. Walters, *History of the League*, p. 350.

35. Quoted in David Davies, *Nearing the Abyss: The Lesson of Ethiopia* (London: Constable and Co., 1936), p. 86.

36. Randolph S. Bourne, *Towards an Enduring Peace* (New York: Charles Scribner's Sons, 1916), p. 310.

37. Matthias Erzberger, *The League of Nations: The Way to the World's Peace*, trans. Bernard Miall (New York: Henry Holt, 1919), pp. 206, 283–297, and Roger Chickering, *Imperial Germany and a World Without War: The Peace Movement and German Society* (Princeton: Princeton University Press, 1975), pp. 114–115.

38. For background, see Miller, *Drafting of the Covenant*, 2:297, 320, 344, 346, 557, and Mee, *The End of Order*, esp. pp. 69, 95.

39. Arthur Walworth, *Wilson and His Peacemakers: American Diplomacy at the Paris Peace Conference, 1919* (New York: W. W. Norton, 1916), pp. 118–119.

40. An acidic contemporary view is C. Howard-Ellis, *The Origins Structure and Workings of the League of Nations* (London: George Allen and Unwin, 1928); for a more clinical postmortem, see Byron Dexter, *The Years of Opportunity: The League of Nations, 1920–1926* (New York: Viking, 1967). Also see Miller, *Drafting of the Covenant*, 2:706–713.

41. Robert Lansing, *The Peace Negotiations: A Personal Narrative* (Boston: Houghton Mifflin, 1921), pp. 288, 290.

III

Blossoms on the Wind: Growth of the International Air Force Concept, 1919–1939

For decades after World War I, its horrors remained fresh in the memories of those who witnessed and endured it, feeding their resolve to avoid a recurrence. The bankruptcy of the balance-of-power politics, the secret diplomacy that brought it about, and military and political leaders' inability to steer the monster on a steady course or bring it to a halt fueled a revolutionary spirit on the political Right and Left. Designers of new military doctrines and technology, like Hitler, Douhet, Mitchell, and Liddell Hart, did not, like appeasers, pacifists and some American isolationists, expect to avoid war altogether but sought to avoid a long and bloody war of attrition like that of 1914–1918. They saw great promise in mechanized warfare providing a quick, cheap way to win battles, a very attractive idea in a world reeling from a war that cost far more and lasted much longer than most reasonable people expected. Outbreaks of mutinies and revolts in most of the combatant nations' armed forces, as well as humanitarian and functional concerns, put a special premium on being able to wage war quickly and decisively before demoralization set in, leading Mussolini to coin his cynical dictum: "War is moral when sudden as a storm."

Dreams of a "new world order" crumbled soon after the Versailles treaty was signed, validating many observers' and actors' doubts at the 1919 peace conference that anything substantial had been done to change the basic structure of the international system that produced the Great War. As tensions eased in the late 1920s, those conflicts looked more like aftershocks of the Great War, leading optimistic "men of good will" to take heart from such snippets of apparent progress toward collective security as the League of Nations' occasional involvement in tiny territorial disputes. In general, however, feeble efforts like the League's dispatching a small peace-keeping force to the border of Colombia and Peru in 1934 underscored the fact that prospects of the League ever drawing the dull, stubby weapon forged at Paris from its sheath in earnest were very slim. Barely a year after the peace treaty was signed, the League Council threw that impotence into bold relief when it dispatched a small international force to oversee a plebiscite in Vilna. Even though the Swiss were hosting the League of Nations headquarters at Geneva, they denied right of transit to French units of the contingent. Although that affair ended without hostilities, a League Council resolution was left in abeyance as the crisis was resolved by direct negotiation between Poland and Lithuania.[1] Another portent of things to come—or more properly, not to come—soon followed, in the Greco-Italian quarrel over Corfu in 1923. This time, the League passed the proverbial buck to the Conference of Ambassadors.

Despite high hopes for "open agreements openly arrived at" created by the Fourteen Points, the design of the League and its undertakings were shrouded in traditional diplomacy. The reversion to treaties, regional blocs, and secret statecraft looked very much like falling off the proverbial wagon after swearing off the morning after and fed the cynical resignation of the "Lost Generation."[2] Many, from Winston Churchill in his memoirs, journalists, and pundits to Roosevelt administration officials, would later blame the League's weakness and even the coming of World War II on the United States' rejecting the Covenant of the League of Nations and its rapid scrapping of much of the weaponry built during its lunge toward Great Power status, 1916–1919. Others assailed the Great Republic for convening the Washington Naval Conference, which scaled back warship construction dramatically, handing an advantage to Japan on the eve of its imperialistic ascendancy. Would greater American involvement in world affairs between the world wars have made all that much difference? That seems very unlikely, if isolationism and pacifism in the United States and appeasement in Western Europe would have evolved much as they did. Defense spending would still have been kept on a very tight leash, and it seems most improbable that even a few American troops would ever have served as the League of Nations' "teeth," let alone on anything close to the scale of much larger U.S. forces that fought under the UN's banner in Korea and the Gulf War. No one else was eager to lead the way along that path, certainly not Britain and France, the major democracies in the League. Rising public hostility to their respective empires and the politically potent pacifism arising from World War I had a blunting effect on their defense establishments much like isolationism did in the United States.

Many condemnations of America's shunning the League were heard during World War II, in the aftermath of the failure of collective security in the late 1930s, which fed a host of firm resolutions that the United Nations organization would keep the peace after the war by maintaining sizable collective military, air, and naval forces. Focusing on American aloofness as a prime cause of subsequent disasters tended to overlook the major wars that continued or erupted in the immediate aftermath of the Treaty of Versailles—in Russia, Poland, the Baltic states, Afghanistan, Morocco, and the Levant. The League stood by as that turbulence produced such dangerous absurdities as Britain and France's backing the losing side in the Greco–Turkish War of 1922 at the same time that they were squaring off over differences concerning the treatment of Germany, even to the point of preparing to exchange bombing attacks. Amid that furor, London's vulnerability led Winston Churchill to describe the British capital, unique among European capitals in its vulnerability to air attack, as a "great fat cow."[3] Such diplomatic stumbling led Sir Maurice Hankey, a key figure in British defense circles, to propose that Britain should unilaterally increase its military capacity to deal with "plainly aggressive actions."[4] In the Imperial Conference later in the year and a subsequent gathering in 1926, the Dominions moved in that direction by applying common standards to British Empire air forces in a plan for a kind of International Air Force, including a "common system of organization and training and. . . . common training manuals, patterns of arms, equipment and stores (with the exception of the types of aircraft), for each part of the Empire as it may determine to co-operate with other parts with the least possible delay and the greatest efficiency."[5]

At the same time, as Raymond O'Connor noted, the U.S. Senate's rejection of the League "deprived" the U.S. Navy "of the role envisaged by Wilson for the enforcement of a world peace." Yet even though America seemed wholly isolated from world affairs, it maintained formal links with the League of Nations through the World Court and International Labor Office, although such slender involvement made isolationists nervous,[6] there were further connections as well. On the other hand, despite major naval arms reductions at the Washington Naval Conference in 1921–1922 and renewal of that agreement in 1930, the U.S. Navy and various vendors confounded President Coolidge's plans for even greater slashes at Geneva in 1927.[7] Nor did lean funding prevent the American fleet from retaining a relatively high level of effectiveness or maintaining a substantial presence in distant waters, from the Asiatic fleet and Chinese garrisons to the "Banana Wars" in Latin America and involvement in the international naval patrol during the Spanish Civil War. Americans who were bent on developing more solid international connections created the Council on Foreign Relations, which, like its British counterpart, the Royal Institute of International Affairs ("Chatham House"), conducted research on long-range issues and made contacts with other internationally minded seekers after peace.

Against that backdrop, the IPF concept emerged in the domain of hopes and ideas from time to time between the world wars, but remained a very faint thread in the tapestry of foreign affairs. In the early 1920s, Weimar Germany's leaders manipulated the West's fear of militant Bolshevism by signing the Pact of Rapallo with the USSR, opening a decade of covert military cooperation and brandishing the specter of a Red Germany. At the same time, German militarists, despite Weimar's liberal constitution and aura of modernity, secretly began rearming in violation of the Versailles treaty. Their way and the Nazis' had been smoothed by the World War I Allies' imposing on Germany a liberal constitution in a way that linked the coming of democracy with the shame and resentment of the 1918 surrender. While Hankey lamented to a colleague that the Americans were "more overbearing and suspicious of us than anyone else" and that "you can't do business with them,"[8] Britain, the major world power at the time, was not eager to forge close links outside the empire. It broke the 1902 alliance with Japan at the Washington Naval Conference and, as noted earlier, opposed French collective security proposals throughout the 1920s and 1930s, aside from the Locarno Treaty, which Britain signed grudgingly in 1925. Designed to ease France's concerns about German resurgence and *Revanchism* by committing local powers to combine forces against potential aggression, that regional agreement moved the center of gravity of western European security even farther away from Geneva.

Despite the marginality of the League, optimists, Americans included, still hoped it might evolve into something stronger.[9] After half a decade of drift, observers could still describe the League as "one of the most firmly established political institutions in the world," which "with or without the United States" would "go forward to larger and larger authority as an international institution."[10] The IPF concept, tenuously linked to the League, was also adrift but still afloat. Despite calls from many quarters for building "a stronger, justly representative" international government with tougher rules for providing "sufficient forces to compel any nation or group of nations to keep the peace,"[11] traditional power politics prevailed, ranging from friction and suspicion among the major democracies, and neutral aloofness, to the tensions arising from the fears of Soviet expansion. But paradoxically, when Leftists came to power in the mother countries of the major empires, they seemed no more eager to abandon territorial holdings than otherwise anticolonialist Americans were willing to foreswear their spheres of influence and the Monroe Doctrine.

The IPF and IAF themes recurred throughout the 1920s, as various members tried to remedy the League of Nations' lack of "a core of organized power." The most notable of those attempts were the Draft Treaty of Mutual Assistance and the 1924 Geneva Protocol. Both failed to jell, but the latter set a high watermark by listing specific steps for putting "teeth" in the League's peace-enforcing mechanism, including the defining of obligations, the setting of trip-wires, and the listing of enforcement procedures, unlike the grandiose and very vague Kellogg–Briand Pact.[12] Initially, the 1924 Protocol appeared to have clear sailing, but after an election in Britain removed the Labourites, Conservatives put their diplomatic weight behind the multilateral, regional Locarno Pact. That portended the course of diplomacy for the next fifteen years. Despite much maneuvering and discussion of the League's architecture, the questions of international security and disarmament were left to traditional bilateral and multilateral diplomacy. As a result, much hard work, including a steady flow of designs for world order structures, came to very little. Some of those schemes were very detailed and carefully thought out, like France's 1930 proposals of a United States of Europe and the Tardieu Plan, which focused on disarmament and the creation of an IPF. Others were broad and unspecific, like those devised by Philip Noel-Baker, the dean of sophisticated British pacifism.[13]

Against that confused backdrop, air power was growing rapidly, despite strictures imposed by widespread disarming after World War I and then by defense economies and the anticipation of formal arms limitation treaties. Despite those constraints, progress in aviation technology continued to accelerate in both qualitative and quantitative terms. As media portrayals of air races and "small wars" presented dramatic images of the results of the rapid advances in engines, fuels, increasingly complex political and diplomatic maneuvers followed the rising curve of development, and production. In the United States, for example, elements in the Democratic Party made common cause with Billy Mitchell, the most ardent American air power proponent, while aviation became a major facet of Italian Fascism's modernist facade. In Germany, the Nazis attracted young air enthusiasts through their paramilitary party organizations, while in Britain Lord Trenchard, as Chief of the Air Staff during the 1920s, built close ties with the Labour Party. At the same time, French parties of various stripes were forging links with various parts of the air services and crafting diverse doctrines. That intricate tapestry was further complicated by a fundamental contradiction—the major western democracies were also imperial powers. If an IPF and/or IAF had been established under international authority at any point between the world wars, it would have bolstered the world order at a time when London,

Paris, Amsterdam, Brussels, Madrid, Rome, Lisbon, Tokyo, Washington, and even Copenhagen were capitals of overseas empires controlled by force or the threat of using it. Again, the basic logic of "what we have, we hold" was visible across most of the political spectrum in empire's mother countries, including the coercive use of military weapons. It was not surprising that an ardent imperialist like Winston Churchill, while a government minister in the early 1920s, helped develop the system of "Air Control" or that members on the Left in the House of Commons often assailed bombing to enforce Britain's imperial writ. It was somewhat perplexing, however, when Lord Thomson, Britain's ill-starred Labour Air Minister in the late 1920s, staunchly defended it.

The democratic-imperial dichotomy engendered further contradictions by forcing mother countries both to maintain armed forces for roles beyond home defense, including major overseas expeditions and colonial skirmishing, and to send many of its best and brightest to rule overseas territories. The hard edge of the system was blurred by idealized visions of imperialism that were then widely accepted throughout the world, even by some among the subject peoples. Practice in the coercive use of air power varied. There was, for example, some differentiation on the basis of race within and among the empires and some inconsistencies as well. Much tighter rules of engagement were applied to the Royal Air Force's operations against white dissidents in Ireland and South Africa after World War I than in more distant reaches of the British Empire, but the Caucasoid Afghans and Kurds also came in for a very solid drubbing, as did Berbers and Druses in the French Empire.

Despite their underlying ethos of force, some defended the British Commonwealth and the Monroe Doctrine as vast "experiments in democracy" engaged in setting "politically backwards people on" their "legs with a minimum of interference from the outside" and claimed that the former was "much more of a working League of Nations than an empire."[14] Others, however, condemned air policing and the IPF/IAF concepts as the raw use of military force to maintain an authoritarian status quo. Conditions on the far edges of empires were rarely detailed in official reports and only glimpsed in media accounts and memoirs, nor were the likely results of an IPF or IAF's applying military power to thwart aggression spelled out in plans and scenarios. The architects of peace-keeping models focused on organization, paying little attention to gradations in their notional applications of force, such as distinctions between intimidation by massing forces at airfields close to a zone of crisis versus making mass flights nearby or directly over an aggressor's territory. Those varied widely, in turn, from preemptive attacks to disrupt, to intervening in a war after hostilities began, when aggressors' and victims' forces were intermingled.

The period between the world wars, like the post–Cold War era, was full of fuzzy grey absolutes in other respects. The great democracies, despite their inconsistencies, including substantial differences in dealing with overseas subjects, were far less brutal at home than the militaristic dictatorships that arose in the 1920s and 1930s. Antiwar sentiment was given a far wider venue in western Europe, Britain, and the United States, and those nations' statesmen were more inclined to search for peace, at least rhetorically. Films, books, pamphlets and posters, debate topics in schools, antimilitarist lobbying, peace ballots and marches, and "No More War" demonstrations all drove home the image of war as an ultimate evil to be avoided at all costs.[15] In the early 1920s, the American isolationist Senator William Borah proposed "outlawing" war, steering around the question of just exactly who would arrest, try, and punish the "outlaws," while later in the decade, American pacifists' lobbying ended "Mobilization Day" and overflights of American cities by military aircraft. In France, mutilated veterans added grim counterpoint to the martial glitter of Bastille Day parades, and in Britain, university students declared their intent to oppose conscription if war should come. Despite idiomatic differences between European and American pacifism, effects on defense expenditures and establishments were similar, leading many politicians and observers who were privately concerned about the rise of the dictators to maintain a pacifist stance in public. In that milieu, a dedicated pacifist with a straight face and little risk of being seriously challenged, proposed entrusting "the whole enterprise of peace to the good faith of the nations who share it" and their "plighted word,"[16] and pacifists of varying stripes in the United States and western Europe condemned the IPF concept as a betrayal of their principles, placing their faith in education and persuasion. So strong were such sentiments that, by the late 1920s, even France's generals, so ferocious at the Paris conference in 1919, had been chained to the strategically defensive doghouse of the Maginot Line.

While critics condemned air control as imperialist bullying or a new form of gunboat diplomacy, IAF proponents often cited it as a precedent,[17] along with the Independent Air Force of World War I. On a slightly divergent track, for example, the military theorist B. H. Liddell Hart, the "hard-headed" pacifist Lord Davies, and J. M. Spaight, a veteran British aviation analyst, offered the latter as a model while arguing in favor of standing League of Nations forces.[18] But such proposals had no practical effect as the wave of pacific optimism crested in the Kellogg–Briand Peace Pact of 1928—the Pact of Paris, whose signatories foreswore war as "an instrument of national policy."[19] The sense of exalted optimism that suffused the drafting and signing of that document was in good part due to what proved to be a brief and transitory moment of international tranquility and unprecedented prosperity in the industrial nations, augmented by Stalin's moderate pose in foreign affairs. As the Soviet dictator brutally consolidated his power, the genial mien of his Foreign Minister, Maxim Litvinov, blended with a foreswearing of the export of the Bolshevik revolution under the rubric "Socialism in One Country." And at that point, even the Fascist dictator Mussolini was widely seen as a creative and dynamic statesman of pacific intent.

That placid seascape soon roughened, as a long procession of crises began that led to World War II. Both the multilateral nature of the Pact of Paris and the site of its signature had reflected a further weakening of the League of Nations and presaged a series of crises and calamities. The Great Depression erupted in America in 1929 and spread to Europe; Japan conquered Manchuria in 1931, then left the League, and Hitler was elected Chancellor in early 1933 soon becoming *Fuehrer*—dictator. In 1935, Italy assaulted Ethiopia, and Hitler left the League and thumbed his nose at the Versailles treaty by creating an air force—the *Luftwaffe*. In 1936, German forces occupied the Rhineland and joined Mussolini in supporting the anti–Republican coup in Spain. Japan invaded China in 1937, and in the late summer of 1938, Hitler, having occupied Austria in further defiance of Versailles, intimidated France and Britain into selling out the Czechs in the Munich Crisis. As those events unfolded, the European democracies reacted sluggishly and confusedly to both the dictatorships' rearmament programs and their aggression. From 1931 until Munich, Britain and France sought to appease the dictators and only then began rearming in earnest, as the United States sought refuge from the mounting threat of war behind a series of Neutrality Acts. Along that descending path, as Nazi Germany, Fascist Italy, and militarist Japan formed what became known as the Axis, they offered the Western democracies some proverbial carrots by claiming that they were protecting Western interests against Communism and chaos. At the same time, the dictators exploited the Western democracies' fear of war by exaggerating the strength of their sizable air forces. They were not alone, however, in overstating the capacity of military aviation. Trenchard in Britain and Billy Mitchell in the United States argued that an air force could be the exclusive instrument of national defense, although many discounted their inflated claims.

While Mitchell's visions of bombing accuracy and stand-off missiles were two generations from operational reality,[20] all architects of IPF and IAF schemes between the world wars shared an exaggerated expectation of air power's potential. Their view of the vitality of the international system was also far over the mark. The two greatest industrial powers, the United States and the USSR, were not League members, and although Russia came to Geneva in the mid-1930s, America remained aloof as various attempts to vitalize the League led down a blind alley. France continued to present ornate designs for a standing League Army, the raising of contingent levies, and the forming of an international police force all to no effect,[21] like Lord Cecil's very muscular plan calling for League members to automatically declare war on any violator of the Covenant.[22] While various models differed in the dimensions and parameters they set on sovereignty and national interest and ranged widely in apparent practicality, it did not matter. Each architect thought their particular blueprint was feasible, but each proved a fantasy. For example, members of the "hard-headed school" of pacifism who dismissed nostrums like the Kellogg–Briand Pact as "mere benevolence" cited the international system of sovereign states as the central evil, a much deeper cause of war than commonly cited factors like economic rivalry and racial and cultural differences.[23] However true that was, they wildly overestimated the likelihood of bringing such forces under control.

Virtually all hues of pacifists, from "purists" to the "hard-headed," tended to confuse their enthusiasm with feasibility.[24] As E. H. Carr observed a generation later, "Rationalism can create a Utopia, but cannot make it real. The liberal democracies scattered throughout the world by the peace settlement of 1919 were the product of abstract theory, stuck no roots in the soil, and quickly shriveled away."[25]

Approaching the drawing board led even the most sober and practical-minded designers of peace enforcement systems to festoon their proposals with such florid lyricisms as "sea patrol,"[26] "Blue Police"—as in policemen of the "far blue,"[27] and "A Navy of Peace."[28] And all the while, the French slogged on, deferring to popular sentiment in the case of the Kellogg–Briand Pact but still striving to put "teeth" in the League of Nations. The plan for an IPF tendered by Edouard Herriot at the outset of the Geneva arms-control talks in November 1932 proposed a multitiered system of military forces, with lightly armed militias providing regional security, and long service contingents maintained by member nations to provide national defense and stand ready to augment the League of Nations' standing units.[29] The latter, a mix of volunteers, like the French and Spanish Foreign Legions serving under League Council control, would provide the first response to aggression—along with an international air force. Herriot proposed removing powerful weapons like tanks, heavy artillery, submarines, battleships, aircraft carriers, and poison gas from individual nations' arsenals and placing them under League control. That led to two special paradoxes. First, the French proposal was opposed by the British delegates who offered the MacDonald Plan, which would allow imperial powers—mainly Britain—to use bombers for "air policing." In clinging to that policy, also known as Air Control, Britain encountered a mirror image of the dilemma it had been presenting to France since 1919. Opponents of the British proposals deemed them products of a concern for national and imperial interests, contrary to effective collective security and disarmament. Air policing, however, was also the product of another form of parochialism—interservice rivalry. From the early 1920s on, the Royal Air Force fended off British Army and Royal Navy attempts to disband it by claiming that Air Control provided the quickest and cheapest way of dealing with insurgents in the far reaches of the empire, leaving details of bombing and strafing out of focus.[30] The ironies were not lost on the Soviet delegates, who responded to the British suggestion that a bomber ban allow an exception for their using air power "in far distant places" by asking if such ground rules would not allow the Red Air Force to bomb London.[31] In respect to the second paradox, some prominent Frenchmen opposed Herriot's plan, while some Britons urged its adoption. Although a few French Leftists endorsed the international army concept and offered armed resistance to Fascism,[32] Léon Blum, soon to be Socialist premier during the *Front Populaire* era, and who, like many others in his party strongly opposed relying on force. Believing that an international armed force could not "exist with the national armies" but could only be formed after their abolition, Blum also feared that if troops carrying the League of Nations' banner were defeated, that might mortally wound Geneva's authority.[33] Blum also shared H. G. Wells' concern that an international army in a disarmed Europe could become a mercenary force, threatening workers' rights by protecting the status quo, and might perhaps establish a repressive regime while claiming to maintain order.[34] Similar anxieties about air power as a potentially revolutionary and coercive force were rippling across Europe. The fear of air forces carrying out *coups d'etat* and establishing dictatorships, reflected in H. G. Wells' *Things to Come,* and Rex Warner's depiction of aerial Bonapartism in his novel *The Aerodrome,* was fanned by the frequent use of aviation in the interwar period as a symbol of national power, like Italo Balbo's mass seaplane flights, Soviet paratroop drops and polar overflights, Britain's Schneider Cup victories, and American long–range bomber flights. In France and Britain, many on the Left feared militarists' using aircraft to stage a coup and dominate the powerless populace, while in the Soviet Union, Stalin negated aerial Bonapartism by gutting Red Army paratroop and heavy bomber forces during the purges of the 1930s. Such concerns were not unique to one side of the ideological spectrum. Both before and after the Nazis came to power in January 1933, Czech statesman Edouard Benes joined German diplomats in opposing French IAF proposals.[35]

During the Geneva arms control talks, military professionals in many countries continued to debate the "roles and missions" of air, ground, and naval forces. During World War I, the main burden of war fighting fell on ground forces and fleets, leaving air power's utility uncertain afterward. It was not clear how much those experiences were relevant, given the rapid pace of technical change and the experience of various "small wars." Until the outbreak of World War II, despite mounting concern about heavy bombers, most military professionals still saw aircraft as properly subordinate to the operations of traditional armies and fleets, and even though short-funding and the rise of pacifism retarded development in some respects, aviation technology continued to evolve quickly enough to substantially increase the durability, speed, size, and range of aircraft, as well as related weapons and communications systems. Despite the marginal effects of mass bombing raids during the First World War,[36] many airmen still shared Douhet, Trenchard, and Mitchell's expectations that air power could win a war without involving armies and fleets and at a low cost in treasure and lives—to the attackers at least. Enthusiasts, however, did not agree on how that might be done. It was not clear, for example, whether many small or a few heavy bombing attacks would yield the best results, or if civilians' morale or civilians outright should be targets. The U.S. Army Air Corps, for example, developed a doctrine based on high–altitude daylight precision bombing of crucial points in enemy nations, despite steady resistance from the Army's ground force elements and contrary to the views of some American airmen, as well as the eminent Italian air power theorist, Giulio Douhet. There was a wide gap between the latter's vision of attacking civilian morale by bombing cities and the doctrine developed at the U.S. Army's Air Tactical School at Maxwell Field in Alabama, based on high-altitude precision raids on critical facilities. While it is not clear how many airmen of that generation read Douhet, they were not his slavish devotees. Some saw using air power for purposes other than supporting armies and fleets as a waste of resources, like Trenchard's successor as head of the Royal Flying Corps, who argued that the British break-through on August 8, 1918, would have been far more successful if Independent Air Force bombers had supported it.[37] Fortunately for the western Allies in World War II, factions that opposed building large long-range heavy bombers gained the upper hand in Japan, Germany, and Italy, keeping air forces harnessed to the tactical needs of the ground forces.

Whatever their particular viewpoint on doctrine and methods, the optimism of air power enthusiasts was contagious. Despite the rise of pacifism between the world wars, glamorous images of combat in the skies became a major theme in popular culture, a trend that began during World War I, when aviators became media stars, and increased throughout the 1920s and 1930s. In 1919, two RAF pilots, Alcock and Brown, first flew the Atlantic as Lindbergh did alone in 1927; air racing and stunting became major spectator sports, and in the 1930s, aviatrixes Amy

Johnson and Jacqueline Cochran set records, and Amelia Earhart disappeared in the far Pacific and became a legend. Against that backdrop, journalists, artists, and filmmakers concocted bizarre blends of fiction and reality, much like pulp magazine writers had during the Wild West era. Gaudy images of aviators and fanciful aircraft appeared on pulp adventure magazine covers,[38] in comic books,[39] children's books,[40] and newspaper comic strips,[41] while movies gave millions a glimpse of flying, something very few could experience at that time. The first Academy Award winning film, *Wings,* was followed by dozens of B–grade films, as well as first–line features like *Hell's Angels, Test Pilot, Dive Bomber, Hell Divers, Dawn Patrol,* and *The Eagle and the Hawk.* Ironically, while many of those films were intended to show the horrors of war, the thrilling images of soaring aloft overrode the underlying message. Serious aviation literature, which reached a much smaller audience, included Faulkner's novel *Pylon* and such lyrical memoirs as Cecil Lewis' *Sagittarius Rising,* Anne Morrow Lindbergh's *Listen the Wind,* and Antoine Saint Exupery's *Wind, Sand and Stars* and *Night Flight.*

In the last half of the 1930s, newsreels and photographs of air war in Ethiopia, China, and Spain stirred fears in many nations of strategic bombing and poison gas being used against major cities at the outbreak of the next major European war. However, some faint and contrary patterns began to appear. In both Spain and China, mass bombing failed to break the will of targeted governments and peoples and attackers suffered unexpectedly heavy losses. Nor did the dictatorships' threatening to use air power wholly cow their adversaries. In the late 1930s, despite the supine posture of the British and French governments, and the American Congress, and the crumbling of the League of Nations, Hitler's and Mussolini's blustering and aggression produced a variant of Newton's Third Law in the form of rising public support for "collective security." A new wave of IPF and IAF plans were much like those that had been drawn since the early 1920s by French and Spanish diplomats. Once again, critics pointed out technical problems and inconsistencies, like the differences between police and armed forces in war. Some noted that the former were neutral third parties who pursued criminals, not other policemen, and others rejected the idea that war in any form could be redemptive, as prisons were meant to be under the correctional logic then popular in the democracies.[42] Such condemnations of relying on force dovetailed with widespread confidence in the attainability of universal disarmament, the utility of peaceful sanctions, and hopes that pacific practices and attitudes were habituating nations to patience and leading them away from violence.[43] Closer to a middle ground but equally fanciful were proposals for Federalist structures without armed forces, based on mutual respect and good intentions.[44]

As the dictators waged war and swaggered menacingly, some pacifists of proven determination, like Lord Allen, permanently disabled while imprisoned during World War I for conscientious objection, were drawn to conclude that a "League with teeth" bore the only solid hope for keeping the peace.[45] British and American suspicions of French motives and the Soviet Union's support for collective security eased slightly when the USSR joined the League, but not enough for them to join Stalin in confronting Hitler over Czechoslovakia. Not until the eve of World War II, with the sky dark and the seas rising, far too late to batten down the metaphorical hatches, did the other major democracies concede that France had been on the right track in trying to structure collective security. The main sticking point had been sovereignty, which Winston Churchill, for all his imperialist ardor, favored limiting. He had supported French proposals for a "League with teeth" at the Paris Peace Conference and, in his history of World War I, presented a Kiplingesque proposal for an International Air Force, extolling Clemenceau's call for creating an international air force in World War I as a revival of the "the old order of chivalry like the Knights Templar . . . to guard civilization against barbarians."[46] Although Churchill would also glide some distance in the early 1920s on the fashionable currents of pacifism while in the Cabinet, he swam against the political tide a decade later by predicting that appeasing the upstart dictators was sacrificing the great democracies' honor to no good purpose. But he was then politically marginal, a back-bencher in Parliament, and a journalist. Few politicians were inclined to join someone with a reputation for romantic rashness and no leverage and take a politically risky stand against the strong tides of pacifism.

From Hitler's seizure of the Rhineland in early 1936 to the outbreak of World War II in late 1939, with Wilson's assurance that America and other nations would aid France against a resurgent Germany echoing hollow, some were willing to take a stand. In the mid-1930s, Hans Wehrberg challenged Blum's arguments and, like some UN Secretaries General a generation later, called for the formation of "at least the nucleus of an international police." He reasoned that a small standing force under League control would have far more practical influence than vast contingents of national armed forces earmarked on paper to respond to unforeseeable and vague contingencies. Repeatedly deploying even a tiny League military unit, Wehrberg suggested, would habituate the world to League policing and intervention. Using contingents from member countries symbolized impermanence and expediency, and their employment would always be vulnerable to power politics, especially in a crisis. Those concerns would resurface in the early years of the United Nations and remain unresolved. Although the UN would far surpass the League in active and successful peace-keeping operations and in deploying armed contingents from various nations, it would also be prevented from forming even a tiny standing constabulary force.[47]

Autopsies of the League began while it was still breathing, and as desperate attempts at resuscitation were under way. Clearly, preventive care had been lacking. In designing the League, much was left intentionally vague in respect to the details of structure of adjudication, peace enforcement, and disarmament. With imperialism validated in the League's mandate system and still viable throughout much of the world, the victorious Allies rejected radical surgery, nor were they inclined to recognize their own very poor health. Full realization of that came slowly, in fits and starts. In 1931, Britain softened its imperialist posture semantically in the Statute of Westminster, which created the British Commonwealth of Nations under an official policy of guiding the colonies to independence, albeit at some unspecified time in the future. That led Field Marshal Smuts, leaving the military, naval, air, and police power underlying the structure out of focus, to describe Britain's dominions as examples of the kind of peaceful enclaves that should be established throughout the world for the benefit of mankind.[48]

Throughout the 1930s, as the evolution of air power became increasingly entwined with mounting crises, the idea of an IAF gained scale and momentum. That seemed unlikely as the arms control talks began at Geneva in early 1932, when hopes of banning an array of powerful weapons were running high. Soon after the conference opened, Japanese planes bombed the Chinese sector of Shanghai, and after the League's investigation and censure, Japan left the organization.[49] Prospects for outlawing bomber aircraft, the main goal of President Hoover, were further dashed in January 1933, when the Nazis came to power in Germany. Soon after Hitler became *Fuehrer*, the last faint sparks of hope for a "League with teeth" faded out in late 1934, after a small multinational force under the League's aegis was deployed in the form of the "Saar Police" to oversee a plebiscite on whether that district should be assigned to Germany or France. (It chose the former.) After League Council members rejected initial proposals for creating a League "Foreign Legion," British Foreign Minister Anthony Eden stepped in, and eventually 1,500 British, 1,300 Italian, 250 Dutch, and 250 Swedish troops went to the Saar, commanded by a British general. With no precedent or provision for "interoperability," the Saar Police's subelements lacked common formats and discipline. After local authorities requested its withdrawal, some observers labeled it a farce, while others, grasping for straws, deemed it a precedent.[50] The only case of military units being deployed under League authority and without a prospect of seeing combat, the Saar Police also paralleled the failure by League peace-keeping missions to resolve the mounting friction between Bolivia and Paraguay that led to the bloodiest conflict in Latin America in the twentieth century—the Gran Chaco War.

Although it is clear in hindsight that the dictators' bluffs and forays during the 1930s undermined any hope of forming an effective international peace maintenance system, that was not so apparent at the time. Indeed, many hoped the sense of threat would heighten the democracies' determination to draw a line on the sand, despite League members' unwillingness to enforce the Covenant's provisions against flagrant aggression in Manchuria and Ethiopia. The crises that dotted the 1930s also highlighted both the special dilemmas of forming a multinational force that might include Italian and German units and the vulnerability of a multinational force if a major League member withdrew or withheld its contingent in the midst of deployment or operations. The powerful inertia of regional security was also dramatized during the Ethiopian crisis when France, for all its tough collective security talk and diplomacy, proved reluctant to enforce sanctions against Italy, a major guarantor of French security under the Locarno Pact. Despite an expanding torrent of proposals for putting "teeth" in the League, Manchuria and Ethiopia made it clear that effective application of armed force at the outset of a major conflict was no more likely than it had been at the Paris Peace Conference, when Wilson tried to soothe the anxious French, and Léon Bourgeois responded with a prediction of the League's demise.[51]

As the arms control talks ground on at Geneva in the mid-1930s and nothing of substance transpired, much was afoot elsewhere. The Italians' heavy use of aerial bombing in conquering Ethiopia in 1935 seemed especially puzzling, since, four years earlier, Italy's delegation at Geneva had proposed scrapping bombers and bolstering international legal protection of civilians against modern military weapons. The images of bombing in Manchuria, Ethiopia, and Spain were conveyed to millions throughout the world by wirephotos and newsreels and made Hitler's leaving the League and creating the *Luftwaffe*—a separate air force—in flat defiance of the Versailles treaty all the more ominous. That triggered an increase in the flow of proposals for creating an IAF and for toughening the League of Nations from many quarters in Europe. In 1936, Japan, now in the firm grip of militaristic adventurers, rejected naval arms limits set by the Washington treaty in 1922 and reaffirmed in London in 1930, as Hitler occupied the Rhineland and announced his intent to rearm while joining Mussolini in aiding the anti-Republican insurgents in Spain. In 1937, Japan's invasion of China included the mass bombing of cities, so that as the Geneva conference ended without result, scenes of urban ruin created by air attacks were regularly featured in the media, dramatizing the fact that the speed, quality, and numbers of military aircraft in service were all increasing steadily. During the five years of the disarmament talks, the speed of bombers in service increased about a third, while that of fighters had doubled.

Although Britain eased its position on the bombing question later in the Geneva conference, it is not clear what would actually have happened if the conferees agreed to ban bombers at the outset. If the course of events in World War II offers any guide, not very much. In that conflict, treaty constraints on resorting to war, attacking civilians, submarine warfare, and using chemical weapons crumbled, although some provisions of interwar agreements were observed in *praxis*, mainly by the Western powers, most notably adherence to the Geneva Convention on prisoners of war and refraining from using gas warfare. The collapse of the initial mutual restraint on air bombing in 1940, however, suggests that a ban agreed to at Geneva would have had little braking effect on the mounting aerial arms race in the late 1930s. Indeed, in the Lewis Carrollian realm of arms control and procurement, a slowdown or halt in production might have led to World War II being waged with more advanced models. In another special irony, Hitler came to power in January 1933, a few weeks after the major powers, including a very reluctant France, had granted Germany equality of rights at Geneva. At that point, the sense that all bets were off led Lord Londonderry, Britain's chief representative, to offer an astounding concession by agreeing with French and Spanish proposals to abolish national airlines and military and naval air forces and internationalize air traffic systems—providing those types of civilian planes not easily converted to military use would not be regulated. At that time, the boundaries between civil and military aviation were less clear than they became after World War II, and the strategic role of commercial aviation was a major source of concern to the great powers, shaping strategic planning, air policy, and diplomacy from the early twentieth century to the Chicago Conference of 1944. (It reappeared during the Cold War from time to time, most dramatically in 1983 when the Soviets shot down a Korean airliner that overflew a military complex in the USSR.) In the 1930s, proposals were made to reduce such anxieties by internationalizing airlines along the lines of postal and telecommunication services, opium control, and the International Labor Office.[52] Those rising apprehensions about the erosion of sovereignty also led some to oppose the idea of forming an IPF outright, while others rejected the concept for practical or moral reasons[53] or saw it as part of a seamless garment requiring disarmament and arbitration mechanisms to make it viable.[54]

As noted earlier, British delegates at Geneva had opposed the Herriot plan and an American proposal to ban bombers to defend their "Air Policing" of restive subjects, mainly in the Middle East and the Northwest Frontier of India.[55] Historians do not agree about how much advantage having the upper or, indeed, the only hand in wielding air power against tribesmen shaped British elites' view of the empire, let alone the shaping of defense policy. In any case, as the Geneva talks began, Prime Minister Stanley Baldwin made his oft-quoted statement that "the bomber will always get through" and asserted that the "only defence is in offense, which means that you have got to kill more women and children more quickly than the enemy, if you wish to save yourself." While some later saw that pronouncement as the origin of the concept of strategic deterrence that flowered in the nuclear age,[56] Baldwin's sense of urgency was not immediately contagious outside defense circles. Despite grisly images in the mass media of the Japanese bombing of Shanghai as the Geneva talks began, like a stockbroker in the first stages of a financial panic, Sir Arthur Salter, a senior international civil servant, along with many other diplomats, dismissed those threatening portents. Like many of his colleagues, Salter would only be moved to register deep concern about the impending catastrophe on the eve of World War II.[57]

In the mid-1930s, the rising threat of German air power led British politicians of various stripes, from Conservative Winston Churchill to Labourite Clement Attlee, to call for creating an International Air Force in Europe to deter aggression and, if necessary, to cover the massing of air contingents from other parts of the world. Each major political party in Parliament offered motions in favor of creating an IPF, although a Liberal resolution in the House of Commons in mid-December 1933 demanding that the government formally support the forming of an International Police Force was withdrawn when the foreign minister cited the arms reductions talks under way as a reason to hold off. A debate on the IPF in the House of Lords in April of the next year was followed by a call from the Woman's National Liberal Federation for an international agreement to abolish military and naval aircraft, internationalize civil aviation, and create an IAF.[58] As clear as the issues seemed on the surface, there was no clear consensus regarding such technical problems as whether "police actions" should be carried out by standing multinational organizations or based on ad hoc "collective security" forces comprised of contingents provided by various nations under either regional auspices or the League of Nations.[59] In 1936, as the skies began to darken, Stanley Baldwin declared on the eve of stepping down as prime minister that "military sanctions" were "an essential part of collective security . . . and cannot be avoided."[60] But they could be, and were, for three more years. Claims by the League's defenders that it had been able to quash an average of two wars a year weighed lightly in the wake of Manchuria and Ethiopia, as mass air bombing in Spain, the Gran Chaco, and China punctuated the last two years of the Geneva talks,[61] which left a sizable diplomatic scrap heap that included President Franklin Delano Roosevelt's (FDR) attempt to carry through the plan of his predecessor Herbert

Hoover for banning bombers,[62] as well as Herriot's scheme, a proposal for open arms inspection, and nearly sixty other designs for creating League forces.[63]

While the high fantasy level in diplomatic and political discourse between the world wars looks bewildering in hindsight and many also thought so at the time, it clarifies why so many in the democracies, elites, and masses alike overlooked or discounted warning signs of what Churchill called "the gathering storm" throughout the 1930s, as well as why they moved so feebly as the symptoms became clear. Historians have marked the League's demise at many different points—the American Senate's rejection of the Covenant; the demise of the Geneva Protocol; Locarno; Manchuria; Japan and Germany's exit from the League; Ethiopia; Spain; China; and, finally, the Munich crisis of 1938, when the great democracies ignored the League as they created their own multilateral mess. Paradoxically, the mounting ripples of crisis and the League's decline led to some convergence of views across the political spectrum. An American pacifist of the hard-headed school, James Shotwell, tacitly recognized the League's stagnation by asserting that regional treaties of mutual assistance were the only realistic solution to the mounting threats from the Axis powers.[64] Despite substantial resistance to the IAF concept on the Left and Stalin's fear of air power, the Soviet dictator moved hundreds of bombers into eastern Czechoslovakia during the Munich crisis in the late summer of 1938 in a vain hope that Britain and France would engage in a joint effort. That was not wholly naive, given the increasing popularity of the IAF concept in Britain in the aftermath of the Spanish Republic's fall to Franco's forces. Although some of that momentum would carry over into the Second World War, it had no practical effect as Europe approached the precipice in the late summer of 1939.

In the United States, President Roosevelt, who usually steered close to public opinion polls, took a guarded position in respect to unfolding events; Congress, even more so.[65] The public, of course, could not ignore them at a time when eighty million movie tickets were sold every week in the United States and tabloid journalism was was at the pinnacle of its influence, regularly featuring grisly images of civilians killed, mutilated, and rendered homeless by modern weapons.[66] American foreign policy was essentially isolationist and neutralist, consisting of such feeble gestures as the Stimson doctrine, which deemed the fruits of conquest illegitimate but provided no means of rendering a balance, and occasional rhetorical sallies against the dictators by FDR.[67] In the realm of popular culture, antiwar "Peaceways" advertisements appeared gratis in high-circulation magazines, and both the "Moral Disarmament" movement and the Ludlow Amendment, a proposal for submitting a declaration of war to a national referendum, gained substantial public support.[68] The rising tide of isolationism, regional security, and pacifism led to the Senate's "Merchants of Death" hearings in 1934, which in turn laid the foundations for a series of Neutrality Acts intended to seal the United States off from involvement in foreign wars. However inadvertently, those laws also signaled Hitler, Mussolini, and Japan that America was determined to stay out of whatever conflict might erupt in Europe or anywhere else outside the Americas, a point underscored further by the U.S. armed forces' "Hemispheric Defense" policy.

A very mild countercurrent to that appeared in the spring of 1935 when, despite Roosevelt's aloofness toward the League, the Senate passed a resolution calling for qualified American membership in the League of Nations and endorsing the spirit of the Pact of Paris and the "fundamental and guiding principles of the Covenant." It also recognized the tension between effective international order and national sovereignty, as well as the Constitution's requiring that Congress approve the involvement of U.S. forces in significant military action, even though, since the early twentieth century, American presidents had frequently deployed troops and warships in punitive and police actions. While Congress permitted wide discretion in such cases, such expeditions were small in scale and closely linked to what government and business elites saw as America's national interests. By late 1936, it was increasingly clear that something big was brewing in Europe, perhaps a recurrence of the First World War with a much higher likelihood of American involvement, despite all the attempts to avoid that. That sense of concern was heightened by the growing range, size, and speed of aircraft, as well as the increase in transoceanic scheduled airlines. While those factors seemed to be shrinking the Americas' strategic moat, the effect was somewhat less dramatic than U.S. Army Air Corps officers and air power enthusiasts suggested as they intensified their "crusade" for an independent air service built around long–range strategic bombing. Throughout the late 1920s and 1930s, airmen staged media events, including endurance flights, sending massed bomber formations to Alaska and Argentina, and the overflight of the Italian liner *Rex* bound for New York by B–17 *Flying Fortress* bombers. While the ship's course and speed were public information, the contingent of news photographers aboard dramatized the point that the interception was far out at sea. Congress, however, resisted such pressures and opposed funding tentative plans for building an intercontinental bomber until after the Geneva arms control talks. Boeing designed and built the B–17 *Flying Fortress* at its own expense, and the Air Corps procured it as a "coast defense" aircraft in 1935, while reorganizing to include an independent strategic bomber force and air staff.

Such attempts to modernize American air power were part of a broader defense buildup amid the shift to neutralism that produced the militant, regional isolationist "Hemispheric Defense" policy driven by growing fears of increasing Nazi and Fascist influence in Latin America, as well as military and naval incursions from outside the Americas. From the sidelines, former President Hoover offered a plan for a "Fortress America," while the armed services and Congress debated over what weapons would best bolster national defense, amid the waves of economic depression that buffeted America from 1929 to 1941. Although Roosevelt was a former Assistant Secretary of the Navy and a naval buff, his interest in air power grew, partly due to the personal magnetism of General Henry H. "Hap" Arnold, chief of the Army Air Corps, 1938–1942, and of the Army Air Forces, 1942–1946, but also because aviation technology was bounding forward. Amid all the doctrinal wrangling, mechanized warfare was demonstrated in the wars in Ethiopia, Spain, China, and to a slightly lesser extent in the Gran Chaco.[69] Hyperbole and publicity, including feature films and staged events, and Nazi, Fascist, and Communist propaganda added to the rising sense of panic, as did book titles like *War over England* and *War on Great Cities*. In the mid-1930s, as Nazi Germany left the League of Nations and began rearming openly in violation of the Versailles treaty, Colonel Charles de Gaulle was urging the French army to mass its sizable armored forces as a mobile deterrent behind the shield of the Maginot Line, while Marshal Henri Petain, then his close colleague but soon to be his adversary, called for increases in France's strategic bombing force.

Across the Channel, the IPF concept came to the fore in public discourse. In London in April 1935, the British League of Nations Union sponsored a conference at Caxton Hall entitled "The Problem of the Air," which included two sessions on "the question of the International Air Force." The focus of the meeting was a Franco–British proposal for a "Western Air Pact" derived from the MacDonald and Herriot Plans at Geneva, which ultimately fizzled. At its core was the internationalizing of civil aviation, then seen as a potential strategic threat because the newer types of commercial transport planes could be converted to bombers, as some soon would be in the Spanish Civil War. At that point in the evolution of aviation technology, several firms were building civil and military planes on essentially the same air-frames, and the contingency of conversion became a major point of discussion at Geneva. At Caxton Hall, speakers' and discussants' comments on the IAF covered the same basic questions and problems that others addressed earlier, including definitions and purposes of force, ethics and principles, the "outworn fetish of national sovereignty," funding, organizational details, authority and control, the role and limits of the League and its Covenant, economic sanctions, zones and bases, disarmament, and civil aviation. Although the basic purposes of an IAF—defense, retaliation, punishment, or deterrence—were not clearly defined, the participants recognized that creating such a force, like the founding of the League, hinged more on collective will among the major powers than on design details.[70]

Italy's invasion of Ethiopia in October 1935 produced fresh enthusiasm for confronting the dictators with some kind of armed international organization. At the same time, one of Britain's most eminent soldiers, Field Marshal Lord Allenby, conqueror of Palestine in World War I and Lawrence of Arabia's commanding officer, spoke out on the issue. Deeply affected by his son's death in the war, he gave a heartfelt speech in 1936 while serving as rector of the University of Edinburgh, arguing that technology, circumstance, and institutions, including a viable World Court, had come into alignment in a way that would allow the establishment of an international system of law and order. An international air force, Allenby suggested, might be "the only means" to deter "an air competition destructive of European civilization." His celebrity gave the speech more gravity than its details,[71] but support for an IAF among "men of affairs" continued to grow. Sir Arthur Henderson, for example, who presided over the Geneva arms control talks, 1932–1937, insisted that the League "must have its police force."[72]

On the other hand, many prominent Europeans clung to their faith that the "spirit of Geneva" would maintain the peace. Despite the League of Nations' marginal influence on world affairs, they insisted, its existence in itself would, over time, gradually strengthen international law, ease tensions stemming from sovereignty and nationalism and, ultimately, lead to a more humane world order, in which aggression would be stifled by the threat of being denied "the right to calculate on the continuance of friendly relations."[73] Such wishful thinking reached its zenith in sophisticated diplomats' view of the League as an ongoing salon in which sophisticated discourse, flavored by wit and wisdom, nudged the nations along to a more civilized way of working out differences.[74] In that ratified atmosphere, even so shrewd and canny a player of hardball politics as Jan Smuts was drawn to claim that "we have found in the League that the conference table is the real weapon."[75] While skeptics labeled "a League in which there is no partial surrender of sovereignty a mockery and a farce,"[76] schemes for reviving it abounded. The designer of a "Contract for Peace or War," for example, envisioned applying legal standards to international relations by creating a Kiplingesque International Control Board with sweeping powers of inspection and the authority to use force while overseeing an immediate halt in naval, aircraft, and munitions production. Armed forces would be reduced by half, reserves frozen at current strength, and large blocs of existing aircraft and ships impounded.[77]

Some Americans hoped to bolster the League by bringing the Great Republic into the League formally. The powerful pacifist and isolationist concerns would be eased by rewriting clauses of the Covenant or circumvented either through subtle congressional maneuvers that expanded existing links with the League or through elegantly vague agreements like the Pact of Paris.[78] On the other side of the issue were advocates of appeasement, some ashamed of the inequities of the Versailles treaty, others fearing Bolshevism or the potential costs and horrors of war or hoping to maintain a balance of power. The result of all those variegated views was a diplomatic crazy quilt, as well as a continuing flow of IPF and IAF designs, perhaps the most idealistic and fanciful of which was the "Peace Army" concept, a uniformed force of volunteers organized along Christian principles of sacrifice and willing to risk martyrdom.[79] When hostilities threatened or erupted, its members would rush to the scene and "put their bodies between the combatants" to prevent an outbreak or halt fighting.[80] The designer of that model took the rise of air power into account, suggesting that land forces had "faded almost into insignificance when compared with the horror of the aeroplane," and admitted that "the 'Peace Army' could hardly prevent combatant nations from carrying out air attacks on each other."[81] Only slightly less fanciful was a proposal to scrap all air forces and stop the building of military planes. Discounting secret production or transfer or aircraft between nations, that scheme was intended to create a long delay in deploying aircraft if war came, assuming that the time required to convert civil aircraft to military purposes or for "specially designed planes . . . [to] be built" would serve as a major brake on hostilities.[82] Especially grandiose was the World Federalists' elaborate plan for immediate worldwide disarmament, after which the nations' military and naval personnel would be placed under a "Universal Policy Authority" to garrison a far-flung "network of Naval Stations, and Bases, and Cruiser Patrols throughout the World" and provide the crews of a "Naval Air Service" and air forces "stationed at points of advantage" to "concentrate in overwhelming force and numbers, wherever danger is threatened."[83]

As crisis followed crisis, architects of peace enforcement systems became increasingly precise and intricate in their draftsmanship, even though the plans ultimately proved as ephemeral as the wispier models of the 1920s. Some observers recognized the gossamer nature of the various designs, like a Quaker peace activist who defined four main categories of peace schemes: *isolationists*, who did not reject military action in every instance; *sanctionists*, who favored using collective military force; *coercionists*, advocates of collective nonmilitary action; and *pacifists*, who were inclined to collective moral action.[84] That tangle of blueprints led the Marquess of Lothian, an articulate spokesman of the hard-headed school, to adjudge pacifists' "wishful thinking" as "more dangerous than that of the hard-boiled realist" who was "solely concerned to avoid war if he can and win if he cannot."[85]

All those dreams crumbled in the aftermath of the Munich crisis of late summer 1938, when Britain and France attempted to prevent the outbreak of war in Europe by negotiating with Hitler. Those meetings took place in the shadow of the apparent lead of the Nazis, Fascists, Japanese militarists, and the Soviets in air strength. As noted earlier, in 1935, Germany created an air force—the *Luftwaffe*—headed by World War I fighter ace Hermann Goering. Although the Versailles treaty forbade Germany's building of military aircraft, the democracies responded cautiously and defensively as the mounting air threat put city dwellers throughout Europe on the front lines of the new form of warfare. The menace of air power came to the fore during the Munich crisis, as the Nazis brandished their apparent superiority in bomber strength. As Prime Ministers Chamberlain and Daladier negotiated a gentlemen's agreement with Hitler, trading Czech frontier territories for the Fuehrer's promise that he would make no more territorial claims in Europe, London's verdant parks were scarred by shelter trenches, and the British government passed out gas masks. Soaring public anxiety was surpassed by that in high places as Air Ministry officials provided Chamberlain with gigantic estimates of casualties from air bombing and gas warfare if war broke out. As Guy le Chambre, French air minister at that time, later claimed that "our aviation was so deficient that we were not free agents."[86] The British experts' forecasts were proved to be over the mark by a factor of ten when war came.

Such gloomy soothsaying, however, was not confined to the corridors of power. On the eve of Munich, a British writer claiming expertise in such matters predicted "a million dead in London" and that the capital would become "the primary battlefield *within a few hours of the outbreak of hostilities*" and the scene of "*a stampede of the threatened populations* beyond all precedent."[87] Another seer came much closer in anticipating "an era still unknown, when the atom will explode under human control . . . for the further destruction of civilization."[88] Although Chamberlain and Daladier were widely condemned, they were supported not only by hard-core appeasers but also by "men of good will" like Franklin Roosevelt, who saw the agreement as a way to buy time as the Allies struggled to catch up with German rearmament. While Chamberlain saw the prospect of war as "a nightmare," he also felt that if "any nation has made up its mind to dominate the world by fear of its force" it should be resisted, for "Under such domination life for people who believe in liberty would not be worth living."[89]

Although FDR held the League of Nations in slight regard, he shared Churchill's view of Hitler as a "bloodthirsty guttersnipe" and veered into the political wind briefly in 1937 in his Chicago Bridge speech, when he called on the democracies to join the United States in taking some kind of collective action under the rubric "quarantine the aggressor." After pacifist and isolationist factions flooded Congress and the White House with angry letters and telegrams, the president kept many of his efforts to oppose the dictators out of public view. He was, however, able to block the Ludlow amendment, and his open expressions of anti–Axis sentiments led some critics to predict he would blunder into war in the early 1940s. FDR's increasing enthusiasm for air power was certainly obvious, especially his call for 50,000 planes a year in early 1941, but despite rising popular interest in the IAF concept, pacifists remained ambivalent about using military power to enforce peace in a "police action."[90]

From Munich to Hitler's invasion of Poland a year later, fresh designs for an international air force appeared as Britain and France began rearming in earnest. While enthusiasts between the wars usually envisioned an IAF as part of an IPF under League control, in support of ground force contingents, a standing constabulary, or an international naval patrol of the kind proposed by Theodore Roosevelt, more and more proposals that appeared in the 1930s envisioned an IAF as the sole element of force, under regional multinational authority rather than the League. Such proposals added to the intensity of debates among the armed services in several countries over the balance of forces. Either creating a standing IAF or earmarking contingents for such an organization would pitch priorities in favor of strategic air forces. Although bomber losses in small wars raised questions about effectiveness, the rapid pace of innovation in aviation technology seemed to be closing the gap between air power enthusiasts' rhetoric and reality. The duel of hyperbole between advocates and skeptics continued into the late 1930s, when airmen in the Western democracies exulted airplanes' increasing capacity to bring force to bear decisively "with extreme rapidity" due to "stupendous" and "tremendous" technical advances,[91] which pacifists' deemed "a manifest return to barbarism."[92]

Despite the high tempo of technical evolution and the growth of major nations' air forces, some opponents of creating an IPF clung to both their confidence in "irresistible moral pressure, diplomatic ostracism and an economic embargo (not blockade or complete boycott)" and the hope that the "transforming of national policies sufficiently" would "make possible the granting of competent jurisdiction to the League of Nations."[93] Despite contrary data from China and Spain, overestimates of air power's potential bloomed in the media and in the corridors of power. In 1938, for instance, senior British officials calculated that the average bombing error was three hundred yards from the aiming point, a number that RAF photo-interpreters would set thirty times higher after two years of bombing operations in World War II.[94] At the same time, a growing split in pacifist ranks between those opposing any form of violence and supporters of collective security was mirrored in tensions among the armed forces in Britain and France as they wrestled with the divergent roles of national defense, a Locarno War, and imperial security. The revulsion for war created by World War I produced a contradiction by feeding both pacifism and what a veteran observer of the League of Nations process identified as the lack of "a real desire to achieve disarmament for its own sake . . . not because nations welcomed budgetary deficits and excessive taxation but because they were afraid"[95]—the theme of Walt Disney's movie cartoon "The Three Little Pigs," which won an Academy Award in 1938.

The mounting fear of the Luftwaffe, which overshadowed the Munich crisis of 1938, was echoed half a decade later by Herbert Morrison, Labour Home Secretary in Churchill's wartime coalition government, when he demanded that "the separate air force, the tommy gun of the international gangster, must be abolished."[96] That was somewhat curious considering the strong links between the Labour Party and the RAF that had been forged by Trenchard while he was Chief of the Air Staff in the 1920s. That rapport became visible during World War II, when the RAF was seen as the least class–ridden and most multinational of British armed services, and some of its celebrities, like Guy Gibson and Don Bennett, were openly contemptuous of the old order. The Left drifted away from its longstanding close relationship with pacifism much earlier, however. In 1935, British Labourites, including Clement Attlee, included an aggressive disarmament plank in their manifesto calling for the disbanding of all national air forces and an end to the arms trade, but the civil war in Spain soon led Labour to support armed collective security, providing that it was aimed at ending imperialism and capitalism in the spirit of the *brigadistas*.[97] Foreign volunteers serving in the International Brigades organized by the Republic to replace the Spanish Foreign Legion, which had gone over to the rebels, the *brigadistas*, whom some saw as a virtual IPF, were not all Communists. The Soviets brought them under firm Stalinist discipline, casting a pall over prospects for "unification of all anti-Fascist forces for a similar struggle in other countries."[98]

While the Republic's fall and vicious infighting among Leftist factions led some to wax fatalistic, others, from pacifists to military professionals and analysts, sought to rally resistance to the rising Fascist tide.[99] Those included Albert Einstein, who, to the consternation of the War Resister's League and other hard-core pacifists, called for stronger collective security measures.[100] New IPF and IAF plans continued to appear, more elaborate and formal in tone and laced with pseudomilitary bureaucratic terms like "An International Strategic Reserve," detailed budgets and subscription formulae, rank tables, rules of engagement, and basing structures,[101] as well as case studies.[102] On the other hand, traditional pacifists held their ground, arguing that "too much reliance" was being "placed upon international law as an agency for peace."[103] Until the Japanese invasion of Manchuria in 1931, Norman Angell, whose steadfastness and idealism had helped him survive his pre–World War I prophecy that no more major wars would occur in Europe, advocated passive options, stressing differences between police and armed forces and denying that force could be an effective instrument of justice.[104] His support of Lord Davies on the issues of collective security and an IPF dramatized the crosswinds that buffeted "men of good will" increasingly throughout the 1930s.[105] A "Peace Ballot" in Britain in 1935 approved the creation of a strong collective security system to confront Mussolini, and a sensation erupted in 1938 when, at the Oxford Union debate, the assembly approved a resolution that young Englishmen would not fight again for king and country. Although widely seen as a statement of adamant pacifism, it actually expressed many students' willingness to fight to maintain peace and counter aggression under international auspices rather than national authority.[106]

As the war in Spain brought about a blending of views on the Right and Left of British politics in regard to collective security, increasing enthusiasm for creating an IPF and IAF was reflected in the growth of the New Commonwealth, a group of "hard-headed" pacifists and opponents of appeasement headed by Winston Churchill, who berated the Baldwin and Chamberlain governments for their feeble response to Hitler's rearming and Mussolini's conquest of Ethiopia.[107] The group's elaborate "peace through police" disarmament program included an international equity court and a police force with a strong air component. Like the Herriot plan, it was designed to remove virtually all heavy, chemical, and technically sophisticated weapons that had appeared since 1914 from nations' arsenals. Under its provisions, twenty-five years after all nations halted technical innovation in their air forces, they would give their planes to an international superauthority, to which members were required give three years' notice before withdrawing. It was not clear how control would be imposed or penalties enforced, beyond the specification that peace was to be maintained in Europe by a tribunal in equity and an international police force under an International Authority, which the United States, USSR, and Japan would also be asked to join. Peace was to be maintained by the International Air Police and an International Maritime Police force, whose members swore oaths of allegiance to the International Authority.[108] IAF recruits would serve five-year active stints and then go into a reserve with provisions for pensions. A formula for levying funds was included, and international infrastructure, administration, and basing locations were spelled out with Palestine, then under the British League of Nations Mandate, as its main facility, designated "a New District of Columbia."

The technically detailed plans of the New Commonwealth, as well as those of other disarmament advocates, contrasted with simpler and bolder goals set by pacifist groups like the Air Disarmament Committee in Britain, who proposed the total abolition of military and naval aircraft.[109] While rising support for an IPF/IAF across the political spectrum led even such resolute pacifists as Beverley Nichols to support such schemes,[110] contrary tides were eroding the assumption on which they were founded—the ability of air power to yield clear, quick results. That became less and less certain in the small wars of the late 1930s, contrary to Lord Davies' assertion on the eve of the Spanish Civil War that "supremacy in the air" had become "the prime factor" in modern war—along with poison gas—and that armies and navies would be left "high and dry" as air fleets battled aloft while the " 'home front' collapsed behind them." Davies expressed confidence that such a cataclysm could be avoided if Europe's twenty-three smaller nations would form a "European Air Police Force," outnumbering five to one the air force of Fascist Italy, then seen as the most assertive dictatorship.[111] But by late 1937, losses inflicted on bomber formations by relatively weak air defenses in Spain and China and the obduracy of those being bombed showed that strategic bombing was less decisive than air power enthusiasts claimed.[112]

Some IPF enthusiasts hedged their bets by suggesting a mix of forces. For example, although Liddell Hart characterized the Independent Air Force of World War I as "a reasonably good augury" of bringing punitive power to bear against an aggressor, he also supported forming a "directly enlisted," standardized international force along the lines of the French Foreign Legion, whose "soldierly spirit" would make it a kind of "martial cult."[113] Some "hard-headed" pacifists saw a standing IPF and/or IAF and economic sanctions as starry-eyed self-delusion in contrast with their goal of "complete international disarmament," to be enforced by a standing, voluntary, multinational force controlled by "a general staff, denationalized, preferably for life," based in strategic denationalized enclaves. They attempted to bypass the major apparent obstacle to such schemes by placing enforcement authority in the hands of an international tribunal without creating an overarching "superstate."[114]

As the figurative diplomatic clock in Europe ticked ever closer to midnight in 1939, even moderates and pacifists were drawn to the IAF concept as a last arrow in the quiver.[115] The international army concept gained support from intellectuals like Leftist celebrity Cyril Joad[116] and British psychologist Pryns Hopkins, who concluded that the cure for such group pathology lay in "an international equity tribunal" exercising control through "a single, democratically-controlled world police-force."[117] Although newsreels of China and Spain provided some graphic details of what "enforcement" might look like, scenes of ruined buildings, frightened or dead civilians, and formations of bombers drove home the image of air power as a relentless force, overshadowing the underlying text that it was not producing a quick, clean decision.

Since the Paris Peace Conference, Britain and other nations, as well as some Americans, viewed the proposals of French diplomats like Herriot, Bourgeois, Tardieu, and Paul–Boncour as maneuvers to bend the League to France's interests and to keep Germany prostrate. The rising threat of war led the British to abandon their longstanding hostility toward such attempts to give the League of Nations "teeth" and led men like Sir Arthur Salter to admit that "the military supremacy of France" had been "the most important single factor in the evolution of the character of the League" and that "the root of French policy was not really an arrogance born of conscious strength, but a sense of insecurity born of the knowledge that her strength was precarious."[118] But that, too, was very late in the game and it made no difference that recurrent crises and rising desperation had brought the IPF and IAF center stage in popular culture.[119] Air war fantasies ranged from the lurid to the visionary,[120] from films like *Wings Over Europe* and *Idiots' Delight* that followed the gloomy format of George Bernard Shaw's *Heartbreak House* by ending in a cataclysmic air attack to the most shocking and realistic of the genre, the 1936 movie based on H. G. Wells' *The Shape of Things to Come.*[121] It portrayed a socialist world state founded in the 1970s under the symbol of a winged disk, whose "Police of the Air and Seaways," headquartered at Basra, maintained "Air and Sea Control" with a force of 3,000 seaplanes flown by "Modern State Aviators." A reflection of the widespread fear of aerial Bonapartism, it showed the air police evolving into a Puritan tyranny after "pacifying" Fascist-Royalists bent on "self-determination." The scenes of massed bomber formations and the hollow ruins of cities, while crude by later special effects standards, produced gasps and nervous laughter in audiences in London.

As air raid siren tests howled in the streets of Europe's cities in the late 1930s, some visions of the IAF moved away from the nuts and bolts of organization and technology and closer to theology. An enthusiast aglow with exalted resolve after France's collapse in June 1940 proposed a "Peace Guard of the World" linked to the "mystical body of Christ,"[122] while some Quakers and others expressed hope that an IPF might be a step toward establishing "a Christian world order."[123] Perhaps the ultimate point in that drift toward the metaphysical was the suggestion by a group of Swedish airmen that "if aerial weapons can provide the means of both imposing and maintaining a system of law applying to all States, the human race has achieved the desired end through the conquest of the air, namely, an advance in the process of evolution."[124] But those in high places were less hopeful. On the eve of World War II, E. H. Carr, a veteran British civil servant, lumped the International Police Force idea with "such purely utopian projects as . . . the Briand-Kellogg [*sic*] Pact and the United States of Europe,"[125] his skepticism very much in tune with Lord Hankey's doubts that America would ever join Britain in "policing and controlling the world."[126]

Despite widespread hope that the Munich Pact had halted the crescendo of crises, the drums beat ever louder in the spring of 1939, when the Spanish Republic fell to Franco's forces, and the *Wehrmacht* broke Hitler's agreement by occupying Czechoslovakia. The major democracies of Europe did not pick up the gauntlet hurled into their faces but stood by and watched as Hitler began to thunder threats at Poland, putting a sharp edge on Madariaga's assertion that "the main utility of a League army would be not its physical but its moral force."[127] Idealism and hopeful expectations tumbled down all at once in every direction, bearing out the conclusion of an analyst looking back at the two-decade history of the League—that "law not backed by force is a methaphysical [*sic*] conception."[128] In the end, there was no trace of collective security in the Anglo–French confrontation with Germany that led to war on September 1, 1939. As the neutrals stayed on the fence like pigeons hypnotized by a circling hawk, the League remained barely alive, and its ghost proved hardier than its living form, haunting those who endured the burdens and horrors of World War II. Geneva's final dull spark, the expulsion of the USSR for assaulting Finland in 1939, was a dying twitch but, ironically, not an indicator of changing attitudes toward international cooperation in the face of aggression or in the prospect of an Axis victory. That led many to conclude that a League with "teeth" or firmer resolve by its members would have prevented the literal holocaust.[129] Millions throughout the world came to believe that peace had been within reach at one point or another from the Paris conference onward and that the last chance had been lost in the 1930s, due to the timidity and irresolution of Britain and/or France and/or America and/or the neutrals and finally, by the USSR. There was certainly enough blame to go around. Traditional diplomacy led to various worthless bilateral and multilateral agreements like Locarno, the Anglo–German naval agreement of 1937, the Munich Pact of 1938, and the Anglo-French-Polish understandings of 1939. But the secret diplomacy condemned in the Fourteen Points produced its most malevolent flower in August 1939 in the form of the Nazi–Soviet Pact.

In those roughening seas, rhetoric substituted for action and substance. An American advocate of armed cooperation's claim that "no collective action . . . by a preponderance of the nations against an aggressor can hereafter be legally called war, or morally considered as war" had no influence on strong American neutralist sentiments.[130] The failure of final efforts, like France's thwarted attempts to create a League of Nations International Army, highlighted the paradox identified during the Paris conference by Charles Evans Hughes—an IPF might be raised only when its use seemed unlikely but could not be used in a real crisis.[131] The fuzziness that enshrouded power politics of that era was epitomized in a proposal made at the International Conference for the Maintenance of Peace in 1937, that in a case of armed aggression, concerned neutral nations should "consult" with the goal of adopting "as neutrals a common and solidary [*sic*] attitude . . . to discourage or prevent the spread or prolongation of hostilities."[132] In those final hours of peace, the IPF concept was frequently cited as a way to keep the peace, and a final attempt to resuscitate the prostrate League was based on organizing a collective security mechanism within its framework.[133] It was too little all along the way from 1919 to 1939, and in the end, too late as well.

Many architects of appeasement—the policy of conciliating Nazi Germany, Fascist Italy, and Japan's militarists—had been infected by what Raymond Aron deemed "catastrophic optimism." But the high costs of their inertia and bewilderment also stiffened the resolve of millions during the Second World War that—in the words of song in an Irving Berlin's show "This is the Army"—"this time will be the last time." Although the League perished, the idea of collective security rose from its ashes even before the war began, when "the chances for a real League" were seen as "dismal enough." As many warmed to the idea of an IPF and/or IAF after hostilities began,[134] such concepts and models became primary themes in British and American public discourse on the postwar world order, and in leaders' and policymakers' rhetoric as well, gaining more popular attention than at any time before or since.

NOTES

1. For details, see Frances Kellor and Antonia Hatvany, *Security Against War*, vol. 1, *International Controversies* (New York: Macmillan, 1924), pp. 254–255.

2. For a vignette of the gloom that affected young, progressive American elites in the spring of 1919, see Arthur M. Schlesinger Jr., *The Age of Roosevelt: The Crisis of the Old Order 1919–1933* (Boston: Houghton Mifflin, 1956), p. 14.

3. For a contemporary reaction, see P.R.C. Groves, *Our Future in the Air* (London: Hutchinson, 1922), pp. 48–49.

4. Stephen Roskill, *Hankey: Man of Secrets*, vol. 2 *1919–1931* (London: Collins, 1972), p. 263.

5. Lionel Curtis, *World War: Its Causes and Cure*, 2nd ed. (London: Oxford University Press, 1945), pp. 47–48.

6. For a sense of atmosphere and constraints, see Frances Perkins, *The Roosevelt I Knew* (New York: Viking Press, 1946), pp. 337–346.

7. Raymond O'Connor, *Perilous Equilibrium* (Lawrence: University Press of Kansas, 1962), p. 126.

8. John F. Naylor, *A Man and an Institution: Sir Maurice Hankey, the Cabinet Secretariat and the Custody of Cabinet Secrecy* (Cambridge: Cambridge University Press, 1984), p. 175.

9. E.g., see Tasker H. Bliss, "World Relations in Their Bearing on International Peace and War," in the Earl of Birkenhead, General Tasker H. Bliss, and Philip Henry Kerr, *Approaches to World Problems* (New Haven: Yale University Press, 1924), pp. 25–71.

10. John H. Clarke, *America and World Peace* (New York: Henry Holt, 1925), p. 133.

11. Oscar Newfang, *The Road to World Peace: A Federation of Nations* (New York: G. P. Putnam's, 1924), p. 64.

12. For a critique of the proposition of levying a percentage of nations' defense budgets for service to the League, see S. C. Vestal, *The Maintenance of Peace or the Foundation of Domestic and International Peace as Deduced from the Study of the History of Nations* (New York: G. P. Putnam's, 1923), p. 456.

13. Philip J. Noel-Baker, *Disarmament* (London: Hogarth Press, 1927).

14. Philip Henry Kerr, "World Problems of Today," Birkenhead et al., eds., *Approaches to World Problems*, p. 104.

15. E.g., see Frederick J. Libby, *War on War: Campaign Textbook* (Washington, D.C.: National Council on the Reduction of Armaments, n.d) [c. 1923].

16. Charles Clayton Morrison, *The Outlawry of War: A Constructive Policy for World Peace* (Chicago: Willett, Clark and Colby, 1927), pp. 18, 178.

17. Lord Thomson, *Air Facts and Problems* (New York: George H. Doran, 1927), pp. 74–75.

18. An analyst of air power from before the First World War, Spaight's views on the subject varied over time; e.g., see his *Aircraft in War* (London: Macmillan, 1914), p. 224; *Pseudo–Security* (London: Longmans Green, 1928), pp. 110–112; *Air Power and the Cities* (London: Longmans Green, 1930), pp. 225–231; *Air Power in the Next War* (London: G. Bles, 1938), pp. 163–166; *Air Power Can Disarm* (London: Pitman and Sons, 1948), pp. 163–169; and *An International Air Force* (London: Gale and Polden, 1932), p. 19.

19. E.g., J. H. Holmes, "Can Pacifists Resort to Force?" *World Tomorrow* 8:4 (April 1925): 101–103, and Esther Everett Lape, ed., *Ways to Peace* (New York: Scribner's, 1924), p. 21.

20. William Mitchell, *Winged Defense* (Washington, D.C.: Kennikat Press, 1971) [orig. publ. 1925], p. 165.

21. "Kepi," "Versailles—Before and After," *Foreign Affairs* 2:2 (December 15, 1923): 207–211.

22. Robert Lansing, *The Peace Negotiations* (Boston: Houghton Mifflin, 1921), p. 247; for an American perspective on demise of the Geneva Protocol, see Harold Josephson, *James T. Shotwell and the Rise of Internationalism in America* (Rutherford, N.J.: Fairleigh Dickinson Press, 1975), pp. 126–128.

23. Philip Henry Kerr and Lionel Curtis, *The Prevention of War* (New Haven: Yale University Press, 1923), pp. 16, 50–64.

24. For an analysis of the influence of interwar pacifist Utopianism on diplomacy and foreign policy, see Maurice Vaisse, "Utopistes at le president," in *Enjeux et Puissances: Pour un histoire des relations internationales au XXe siecle* (Paris: Publications de la Sorbonne, 1986), pp. 175–181.

25. Edward Hallett Carr, *The Twenty Years Crisis: An Introduction to the Study of International Relations* (London: Macmillan, 1962), p. 27.

26. Salvador de Madariaga, *Disarmament* (New York: Coward-McCann, 1929), p. 140.

27. Ollivant, *Next Step*, pp. 111, 120.

28. Nicholas Murray Butler, *The Path to Peace: Essays and Addresses on Peace and Its Making* (New York: Charles Scribner's Sons, 1930), p. 156; also see Philip C. Jessup, in *American Neutrality and International Police* 11:3 [World Peace Information Pamphlet] (Boston: World Peace Foundation, 1928).

29. For French perspectives on interwar diplomacy relative to disarmament and air power, see Maurice Vaisse, "Le proces de l'aviation de bombardement," in "L'Aviation Militaire Francaise 1919–1939," special issue of *Revue Historiques des Armees* no. 2 (1977), pp. 41–61; for background on the first phase of the Geneva talks, see Maurice Vaisse, *Securite d'Abord: La politique francaise en matiere de desarmament, 9 decembre 1930–17 avril 1934* (Paris: Editions Pedone, 1981), pp. 200–222, 292–323, 378–389.

30. E. J. Kingston-McLoughry, "Policing by Air," *Winged Warfare: Air Problems of Peace and War* (London: Jonathan Cape, 1937), pp. 201–257.

31. See Arthur Henderson, *Preliminary Report on Work of the Conference* [for the Reduction and Limitation of Armaments] (Geneva: League of Nations, 1936), pp. 17, 88–89.

32. E.g., Lucian le Foyer and Frederic Longuet.

33. Léon Blum, *Peace and Disarmament*, trans. Alexander Werth (London: Jonathan Cape, 1932), pp. 137–138.

34. Ibid., pp. 142–143.

35. For Benes' view, see Memorandum from Pierre Cot, then Undersecretary of State for Foreign Affairs to M. Paul-Boncour, Secretary of State, January 25, 1933, on a conversation with Edouard Benes, in *Documents Defense Francaise*, 1st series, vol. 2, no. 238; for a German view on the eve of the 1932 Geneva arms control talks, see Max Graf Montgelas, "Internationale Truppenkorper, gemeinschaftsliches Vorgehen der franzosische Vorschlag," *Zeitschrift fur Politik* 21 (1932): 857–865.

36. Douglas Robinson, *The Zeppelin in Combat* (London: G. T. Foulis, 1962), p. 308.

37. Andrew Boyle, *Trenchard: Man of Vision* (New York: Norton, 1962), p. 308.

38. E.g., *G–8 and His Battle Aces*.

39. E.g., *Airboy, Blackhawk, Sky Man*, and *Enemy Ace*.

40. E.g., the *Dave Dawson* and *Tom Swift* series.

41. E.g., *Barney Baxter, Captain Midnight and the Secret Squadron, Connie, Hop Harrigan, Scorchy Smith, Smilin' Jack, Flyin' Jenny, Bruce Gentry, Johnny Hazard, Steve Canyon, Terry and the Pirates*, and *Tailspin Tommy*. Several of these also appeared in comic books, and/or radio or television broadcasts. For a contemporaneous perspective, see Birt Darling, "Aviation in the Comics," *Aero Review* 5:1 (June 1946): 25–29.

42. Sherwood Eddy and Kirby Page, *The Abolition of War* (New York: George Doran, 1924), pp. 108–110. Exceptions included France's Devil's Island, Britain's Andaman Island penal colony, and America's Alcatraz.

43. E.g., the Esher Plan, "Resolution XIV" and the Draft Treaty of Mutual Assistance, and the Geneva Protocol of 1924, all of which came to naught.

44. One such grand vision was Max Green, *The Problem of Universal Peace and a Key to Its Solution* (Philadelphia: George H. Buchanan, 1924), p. 14.

45. Thomas C. Kennedy, "R. Clifford Allen," in Harold Josephson, et al., eds., *Biographical Dictionary of Modern Peace Leaders* (Westport, Conn.: Greenwood Press, 1985), pp. 17–19.

46. Winston S. Churchill, *The World Crisis: The Aftermath* (London: Thornton Butterworth, 1922), pp. 7–12, 25–27.

47. Hans Wehrberg, *Theory and Practice of International Policing* (London: Constable, 1935), p. 93; for a contemporaneous perspective on the UN force dilemma, see Andrew Martin, *Collective Security: A Progress Report* (New York: UNESCO, 1951), pp. 166–168.

48. Jan Christiaan Smuts, "The British Empire and World Peace," *Journal of the Royal Institute of International Affairs* 9:2 (March 1930): 141–153.

49. For a reaction of the moment by a champion of using force to maintain the peace, see Lord [David] Davies, "Force and the Future," *Force* (London: Ernest Benn, 1934), pp. 156–209, esp. p. 163.

50. E.g., Cedric Fowler, "The First World Army," *New Outlook* 165:3 (March 1935): 17–21.

51. See "An International Air Force," *Flight* 28:1405 (November 28, 1935): 549.

52. George Lansbury, *My Pilgrimage for Peace and Peace Through Economic Cooperation* (New York: Garland, 1972), p. 260; also see Kenneth W. Colgrove, *International Control of Aviation* (Boston: World Peace Foundation, 1930); W. Arnold Forster, "Disarmament and Security," in Gilbert Murray, et al., eds., *Problems of Peace*, 8th Series (London: George Allen and Unwin, 1933), pp. 565–567; for British official perspectives, see PRO AVIA 1834, *Operation of Air Route from Marseilles to Constantinople or Syria by French Company* [1925]; PRO AVIA 1878, *Series of Flights to be Organized during 1930 by French Company to Indochina;* and PRO AIR 2/664 Report #13, March 3, 1933, on *Internationalization of Civil Aviation and Constitution of an International Air Police;* and for an American journalistic view, see "The Coming Struggle for the Airplanes," *Fortune* 23:39 (March 1941): 124 ff. A bibliographical overview is Wybo P. Heere, *International Bibliography of Air Law, 1900–1971* (Leyden, Holland: A. W. Sijthoff, 1972).

53. E.g., A. A. Milne, *Peace With Honor* (New York: E. P. Dutton, 1935), pp. 211–212, and Norman Thomas, *War: No Glory, No Profit, No Need* (New York: Florence A. Stokes, 1935), pp. 182–184.

54. See Benjamin F. Trueblood, *The Development of the Peace Idea and Other Essays* (Boston: Plimpton Press, 1932), pp. 148–149.

55. P.R.C. Groves, *Behind the Smoke Screen* (London: Faber and Faber, 1934), p. 315. Other contemporary perspectives are J.H.H. [pseudonym], " 'Police' Bombing in Outlying Regions," *The Naval Review* 21:3 (August 1933): 461–465; and Oliver Lyman Spaulding, *Ahriman: A Study in Bombardment* (Boston: World Peace Foundation, 1939), pp. 102–103. A contrary view from a senior airman is former Chief of the Air Staff and Controller General of Civil Aviation Sir Frederick Sykes, *Aviation in Peace and War* (London: Edward Arnold, 1922), esp. p. 132. A recent scholarly analysis is Charles Townsend, "Civilization and 'Frightfulness': Air Control in the Middle East Between the Wars," in Chris Wrigley, ed., *Warfare, Diplomacy and Politics: Essays in Honour of A.J.P. Taylor* (London: Hamish

Hamilton, 1986), pp. 142–161. Also see Elmer Berman Scoville, "The Royal Air Force, the Middle East, and Disarmament 1919–1934" (Ph.D. Dissertation: Michigan State University, 1972). Among many pertinent records in the British Public Record Office are AIR 5/168 [CD 21], "The Power of the Air Force and the Application of This Power Hold and Police Mesopotamia," June 1921; and AIR 5/1325, "Paper on the Use of the RAF on the Northwest Frontier of India," 1935. For a discussion of the influence of the Geneva talks on interservice rivalry in Britain in the mid–1930s, see AIR 5235, L. Darvell, staff paper, October 1935, para. 11. For somewhat divergent views, see the works of David Omissi, and Philip S. Meilinger, "Clipping the Bomber's Wings: The Geneva Disarmament Conference and the Royal Air Force, 1932–1934," *War in History* 6:3 (July 1999): 306–330, esp. p. 311.

56. Speech in the House of Commons, November 10, 1932.

57. Sir Arthur Salter, in *The United States of Europe and Other Papers* (Freeport, N.Y.: Books for Libraries, 1970) [orig. publ. 1933].

58. For details, see *Liberal Magazine* 42 (January 1934), pp. 39; and 42 (June 1934): 169.

59. Clyde Eagleton, *Analysis of the Problems of War* (New York: Ronald Press, 1937), p. 123.

60. Quoted in David Davies, *Nearing the Abyss: The Lesson of Ethiopia* (London: Constable and Co., 1936), p. 83.

61. A caustic contemporary account is John Wheeler–Bennett, *The Pipe Dream of Peace: The Story of the Collapse of Disarmament* (New York: William Morrow, 1935).

62. American policy statements made no mention of using military force to maintain peace.

63. "Through the Jungle to Peace," *Literary Digest* 112:8 (February 20, 1932): 16. For background on the diplomacy of the collective security wrangling, see Francis de Tarr, *The French Radical Party: from Herriot to Mendes-France* (London: Oxford University Press, 1961).

64. James Shotwell, *On the Rim of the Abyss* (New York: Macmillan, 1936), p. 179.

65. See Philip C. Jessup, *International Security: the American Role in Collective Action for Peace* (New York: Council on Foreign Relations, 1935) for a moderate, formalistic exposition of the U.S. position in the mid-1930s.

66. International Parliamentary Union, *What Would be the Character of a New War?* (London: P. S. King and Son, 1937), esp. pp. 75, 94–95.

67. Named for its architect, Secretary of State Henry Stimson; for perspective, see Frank H. Simonds and Brooks Emery, *The Great Powers in World Politics* (New York: American Book Co., 1939), pp. 588–590.

68. For details, see Arthur Henderson, *Preliminary Report on the Conference for the Repeal and Limitation of Arms* (Geneva: League of Nations, 1936), pp. 143–144.

69. Harold Butler, *The Lost Peace: A Personal Impression* (London: Faber and Faber, 1941), p. 174.

70. "The Question of an International Air Force," in *The Problem of the Air: Being the Report of a Conference on Aviation as an International Problem, Held at the Caxton Hall, Westminster, at the Invitation of the League of Nations Union, on April 3rd and 4th 1935* (London: League of Nations Union, c. 1935), pp. 30–44.

71. Field Marshal Lord Allenby, "World Police for World Peace," *International Conciliation* 313: 10 (October 1936): 297, 484; for some student reaction at the time, see "The Rectorial Address," *The Student (Edinburgh University Magazine)* 32 (1936): 276–277.

72. Quoted in W. Bryn Thomas, *An International Police Force* (London: Allenson and Co., 1936), p. 9.

73. James T. Shotwell, *Was as an Instrument of National Policy and Its Renunciation in the Pact of Paris* (New York: Harcourt Brace, 1929), pp. vi, 221, 266–267.

74. E.g., see A. C. Temperley, *The Whispering Gallery of Europe* (London: Collins, 1938).

75. Smuts, "British Empire and World Peace," p. 150.

76. Bryn Thomas, *An International Police Force*, p. 117.

77. Stephen Raushenbush and Joan Raushenbush, *The Final Choice: America Between Europe and Asia* (New York: Reynal and Hitchcock, 1937), pp. 282–287.

78. Most notable of such attempts were the Davis Declaration at the Disarmament Conference of 1932, which proved irrelevant when the conference failed, and the long-debated but abortive Pope Resolution of 1935. A contemporaneous perspective is Livingston Hartley, *Is America Afraid?* (New York: Prentice-Hall, 1937), pp. 377–381, 389, 413–416.

79. Not to be confused with the "Peace Army" (Shanti Sena) in India, created in the late 1950s by Vinoba Bhavi and headed by Narayan Desai.

80. Henry Brinton, *Peace Army* (London: Williams and Norgate, 1932), pp. 71–78.

81. Ibid., p. 75.

82. Ibid., p. 77.

83. Herbert John Painton, *How War Was Abolished: Plans for World-Wide Peace: A Prophecy of the Near Future* (London: Simpkin, Marshall, Hamilton, Kent and Co., 1924), pp. 119–121, 139.

84. A. Ruth Fry, "An International Force?" in *An International Force?* (London: Edgar G. Dunstan, 1934), pp. 6–9.

85. Marquess of Lothian, *Pacifism is Not Enough Nor Patriotism Either* (Oxford: Clarendon Press, 1935), p. 7.

86. Quoted in Edward Mead Earle, "The Influence of Air Power Upon History," *Yale Review* 35:4 (June 1946): 586.

87. Author's emphasis; Frank Morison, *War on Great Cities: A Study of the Facts* (London: Faber and Faber, 1937), pp. 184, 187. Also see Davies, "Force and the Future," pp. 156–209, esp. p. 163.

88. Elvira Fradkin, *Air Menace and the Answer* (New York: Macmillan, 1934), p. 273.

89. Henry M. Wriston, *Prepare for Peace!* (New York: Harper and Brothers, 1941), p. 40.

90. George Malcolm Stratton, *Social Psychology of International Conduct* (New York: D. Appleton, 1929), p. 344.

91. William McDougall, *Janus: The Conquest of War: A Psychological Inquiry* (New York: E. P. Dutton, 1927), p. 142.

92. "Neon," *The Great Delusion: A Study of Aircraft in Peace and War* (New York: Dial, 1927), p. xxxvi.

93. Sherwood Eddy and Kirby Page, *What Shall We Do About War?* (New York: Eddy and Page, n.d., c. 1930), p. 69.

94. PRO AIR 2/8812, "Bombing Policy Subcommittee Minutes," March 22, 1938.

95. Temperley, *Whispering Gallery of Europe*, pp. 71–72.

96. Victor Altman, *The International Police and World Security* (London: Alliance Press Ltd., 1945), p. 9.

97. H. N. Brailsford, "An International Police," *World Tomorrow* 17:2 (January 18, 1934): 32–33; and Patrick Kyles, *Covenant Without the Sword: Public Opinion and British Defence Policy* (Waterloo, Ont.: Wilfred Laurier Press, 1983), p. 173.

98. R. Dan Richardson, *Comintern Army: The International Brigades and the Spanish Civil War* (Lexington: University of Kentucky Press, 1982), pp. 91–92.

99. E.g., "The human mind turns out not to have progressed, but actually to have gone backwards in its discovery of anarchy as the last word in international politics," Esme Wingfield-Stratford, *They That Take the Sword* (New York: William Morrow, 1936), pp. 399–400.

100. Otto Nathan and Heinz Norden, *Einstein on Peace* (New York: Simon and Schuster, 1960), pp. 223–240.

101. E.g., *Information Papers No. 17: The Character of International Sanctions and Their Applications* (London: Royal Institute of International Affairs, 1934); L.E.O. Charlton, *The Menace of the Clouds* (London: William Hodge, 1937), esp. pp. 205–293.

102. E.g., W. Ross, "An International Police Campaign," [Seymour Expeditions, 1900] *Fighting Forces* 12:2 (June 1935): 179–181.

103. E.g., John Francis Kane, *War or Peace? A Forecast* (New York: Timely Books, 1936), p. 34.

104. Norman Angell, *Peace and the Plain Man* (New York: Harper and Brothers, 1935), p. 163.

105. Lewis Busceglia, "Norman Angell," in Harold Josephson, et al., eds., *Biographical Dictionary of Modern Peace Leaders* (Westport, Conn.: Greenwood Press, 1985), p. 29.

106. R. B. Mowat, *Anarchy or World Order* [Problems of Peace, Tenth Series] (New York: Oxford University Press, 1936), p. 168; for a survey of pacifist thought with detailing on the the 1930s, see David A. Martin, *Pacifism: An Historical and Sociological Study* (New York: Schocken Books, 1966).

107. Naylor, *A Man and an Institution*, pp. 254–256.

108. R. E. Hopwood, "A Proposal for an International Maritime Police," *New Commonwealth Quarterly* 1 (1935–1986): 129–138.

109. Its pamphlet series included *Disarm the Air*! (London: Air Disarmament Committee, 1935), *A.B.C. of Air Disarmament* (London: Air Disarmament Committee, 1935); and *The Air Menace and How to End It* (London: Air Disarmament Committee, 1937) of a pamphlet, *The Air Menace* (London: Air Disarmament Committee, 1938). For its basic position, see George A. Innes [ADC General Secretary], "The Government and the 'Bomber,' " in *Peace*, March 1938, p. 185.

110. Beverley Nichols, *Cry Havoc!* (Garden City, N.Y.: Doubleday, Doran, 1933), pp. 260–261.

111. Davies, *Nearing the Abyss*, pp. 102–103, 116–117.

112. E.g., Spaulding, *Ahriman*, pp. 53–55, 64, 85. An editorial in *Flight* noted mass bombing's failure to break "the spirit of any nation or party, and warned that "brutality produces anger" rather than "submission" or "a short cut to national victory," "Useless Barbarity," *Flight* 32:1500 (September 30, 1937): 326; also see "Bombs and Bombs," *Flight* 32:1510 (December 2, 1937): 532.

113. B. H. Liddell Hart, "International Force," *International Affairs* (March 1933): 205–218.

114. J. J. van der Leeuw, *Why a World Police is Inevitable* (London: New Commonwealth, 1937) [orig. publ. 1934], pp. 14–15.

115. See Storm Jameson, *Challenge to Death* (London: Constable and Co., 1934), esp. Philip Noel-Baker, "The International Air Police Force," pp. 206–239.

116. C.E.M. Joad, *Why War?* (Harmondsworth, England: Penguin, 1939), pp. 222–225; also see Martin Ceadel, *Pacifism in Britain, 1914–1945: The Defining of a Faith* (Oxford: Clarendon Press, 1980), pp. 139–140.

117. Pryns Hopkins, *The Psychology of Social Movements: A Psycho-Analytic View of Society* (London: George Allen and Unwin, 1938), p. 116.

118. Arthur Salter, *Security: Can We Retrieve It?* (New York: Reynal and Hitchcock, 1939), pp. 116–117, 126–127.

119. H. Rowan Robinson, *Sanctions Begone! A Plea and a Plan for the Defense of the League* (London: William Clowes, 1936), esp. pp. 55–62.

120. For an overview of the period, see Lee Kennett, *A History of Strategic Bombing* (New York: Charles Scribner's Sons, 1982), pp. 67–68, and Thomas, *An International Police Force.*

121. H. G. Wells, *The Shape of Things to Come* (New York: Macmillan, 1933).

122. A. R. Bandini, "Give Peace a Sword," *Catholic World* 152:97 (October 1940): 52.

123. Ceadel, *Pacifism in Britain*, pp. 158–159.

124. A. K. Bratt and G.B.R. Sergel, "Aerial Weapons and Future Wars," in Norman Angell, et al., eds., *What Would Be the Character of a New War? Enquiry Organized by the Inter–parliamentary Union in Geneva* (London: Victor Gollancz, 1933), p. 95.

125. Carr, *The Twenty Years' Crisis*, p. 20. An overview of that line of diplomacy is P. E. Corbett, *Post-War Worlds* (New York: Farrar and Rinehart, 1942).

126. Stephen Roskill, *Hankey: Man of Secrets*, vol. 3 *1931–1963* (London: Collins, 1974), p. 524.

127. Salvador de Madariaga, *World's Design* (London: George Allen and Unwin, 1938), p. 188.

128. S. Engel, *League Reform: An Analysis of Official Proposals and Discussions 1936–1939*, vol. 11 (Geneva: Geneva Research Center, 1940), pp. 3–4.

129. E.g., see James Avery Royce, *Broken Star: The Story of the League of Nations, 1919–1939* (Swansea: Christopher Dawson, 1978), p. 197.

130. Denna Fleming, *The United States and World Organization, 1920–1993* (New York: Columbia University Press, 1938), p. 536.

131. F. P. Walters, *A History of the League of Nations* (London: Oxford University Press, 1952), p. 444.

132. "The International Conference for the Maintenance of Peace," *International Conciliation* No. 328 (March 1937): 228.

133. See Engel, *League Reform*, pp. 31–32.

134. Stuart Chase, "A Policeman for the Planet," *The New Western Front* (New York: Harcourt Brace, 1939), p. 125.

IV

In Fullest Bloom: Manifold Visions of an International Air Force in World War II

Soon after the Wehrmacht assailed Poland early on the morning of September 3, 1939, images of air power began to flow across the world from the battle zone. While the flood of wirephotos and newsreels showed the rapid tempo of what reporters deemed the *blitzkrieg* (lightning war), it did not shock those who watched the Nazis test their new techniques of warfare in Spain. In the weeks that followed, millions in Europe and elsewhere were pleasantly surprised that no massive bombing raids or poison gas attacks were carried out in the opening months of World War II in Europe, aside from the Luftwaffe's bombing of Warsaw as a "defended place." Britain, France, and Germany acceded to President Roosevelt's call for restraint, in good part because none were eager to take the onus for being the first to attack cities. The French held back completely, as reluctant to trigger a retaliatory spiral as they were during World War I, while the Germans and the British bombed some military and naval targets away from populated areas, like fleet anchorages.[1] Nor did the Wehrmacht attack cities when it launched its major offensives in the West, first in Scandinavia in April 1940, and then against France and the Low Countries in May. The Nazis attributed the well-publicized and widely condemned raid on Rotterdam in May 1940 that killed several hundred Dutch civilians to a command-and-control glitch, and it may well have been true, but the furor it stirred up made it clear that the dogs of unrestricted aerial warfare were tugging hard at their leashes.

The gloves first came off in the brief "Winter War" of 1939–1940, as the Soviet air forces killed between 600 to 2,000 Finnish civilians in over two thousand bombing raids. There, the pattern of Spain and China was repeated in microcosm. Most civilians being bombed did not crack, and beyond Russia's loss of over 300 planes was the diplomatic cost: the USSR was expelled from the League of Nations in its twilight hours.[2] The main combatants' restraints on aerial warfare lasted until late August 1940, when British jamming of German radio-navigation guidance beams caused a Luftwaffe unit to drop bombs on a London suburb accidentally. The British War Cabinet then ordered a reprisal raid on Berlin to relieve the Luftwaffe's heavy pressure on their fighter bases, an escalation that reverberated throughout World War II in Europe, and later, in the Far East, where Japanese bombs had already killed tens of thousands of Chinese civilians prior to the outbreak of World War II in Europe. While the German attacks on British cities in 1940–1941 were a major bound forward in scale, they fell far short of the level of cataclysm anticipated before the war. Although the "Blitz" killed almost sixty thousand, Chamberlain's advisors during the Munich crisis of 1938 had offered him projections of four times that number in the first six months. As the aerial Battle of Britain raged, the Germans abandoned daylight bombing for night attacks, but as in China, Spain, and Finland, they failed to break the morale of those being attacked, while the raiders suffered heavy losses. After a renewal in the spring, the Blitz eased off when the Nazis invaded Russia in June 1941.

Despite that confounding of enthusiasts' hopes for a quick, clear decision that would demonstrate the supremacy of air power, both the Luftwaffe's stunning successes in the blitzkrieg campaigns and the Royal Air Force's victory in the Battle of Britain made it clear to all but a handful of military and naval traditionalists that aircraft were now a vital element in waging war. Nazi ground forces marched from triumph to triumph under the umbrella of air supremacy in the Balkans, Russia, and North Africa in 1941, as the Japanese did during the opening campaigns of the Pacific War in early 1942, generating waves of anguish on the one hand and further hyperbole and hyperoptimism on the other. Many concluded that Britain, France, the armed neutrals, and/or the USSR had thrown away a significant opportunity in the mid-1930s, when the balance was in their favor, by failing to mass their air forces against Germany, Italy, and Japan. Air power advocates like Alexander P. de Seversky echoed Billy Mitchell in urging that America's defense buildup be concentrated overwhelmingly on building a fleet of long-range heavy bombers.

When war began in Europe, despite strong neutralist sentiments in the United States, an official defense policy of "Hemispheric Defense," and substantial enthusiasm for former President Herbert Hoover's "Fortress America" concept, strong currents were also already running in the direction of internationalism. A Gallup poll showed half of Americans surveyed favoring an International Police Force,[3] and after the Japanese Pearl Harbor attack brought the United States into the war in December 1941, challenges to the principle of national sovereignty appeared in the popular press and in public policy analysis, much as they had during World War I. Throughout the war, both the IPF and IAF concepts gained major popular acceptance, as many who attributed the coming of World War II to the League of Nation's weakness called for a new version of it with "teeth."[4] Despite a wide gap between policymaking elites and the general public on that question,[5] the complex problems of enforcing peace after the war had come under close analysis in the U.S. and British governments' upper echelons before Pearl Harbor. Secretary of State Cordell Hull formed the 100-member Group to Study the Organization of the Peace to advise the State Department,[6] while in Britain, discussions over an international peace-enforcing mechanism also intensified, despite the bleak prospects for victory after Dunkirk, and France's surrender. The New Commonwealth continued to press for an "International Air Police," but left some inconsistencies unresolved, like the contradiction between the definition of an IAF as being "distinct from a military force" and its need to produce "a deterrent effect."[7] Some proponents offered the convoluted argument that an IAF would excite less passion in opponents across national boundaries but could stir strong feelings among "an opposing political faction or group" within a nation participating in collective action,[8] since it might very well wage "guerre totale," including the bombing of cities. One justification of the anticipated casualties took the form of an analogy between such massive losses and the deaths of innocent people caused by police and military forces when they suppressed riots.[9]

When the Luftwaffe mounted its air "blitz" of Britain in the summer of 1940, such fine distinctions seemed irrelevant to those under air attack. Critiques of strategic bombing appeared in many quarters, official and unofficial, in public and in secret, and not all Britons who were bombed favored retaliating. One observer lamented that "the aeroplane" had introduced "a continuous threat of violence; at least a play on the emotion of fear in a war of nerves, at most a discriminate bombardment of the homes of the people."[10] A Canadian analyst, also cool to the idea, conceded that an international "air force would probably be the easiest to set up because of the scientific nature of its equipment and the lack of national prejudice among its personnel." Citing the rapid pace of technical advance in aviation, he warned that even though air power appeared to be a cheaper and more mobile weapon than other ways of waging war, substantial costs might lie hidden in the rolling waves of obsolescence that arose from such advances. Rising waves of innovation constantly rendered a sizable portion of all air forces inferior to their most recently delivered equipment—and their enemies equipment.[11] At the same time, mounting uncertainties regarding the effectiveness and the morality of strategic bombing were being thrown out of focus by the Wehrmacht's stunning victories in the spring of 1940, rekindling old arguments that some viable form of international government must precede "compulsory arbitration, collective security, and an international political force." If that sounded very fanciful,[12] the seemingly invincible Nazis sometimes struck a pose of moderation as they used propaganda and diplomacy to exploit their European adversaries, vassals, and neutral nations' reverence for sovereignty. In capitalizing on longstanding hostilities there and in the United States, the Nazis cadged a free ride on the same attitudes that had thwarted the forming of collective security arrangements between the wars. Both the USSR and the United States remained neutral through the first third of World War II, the former providing the Nazis aid, comfort, and materiel, while America's "measures short of war" supported Britain, France—and the USSR after the Nazi invasion. While the "Hemispheric Defense" and "Fortress America" rubrics reflected many Americans' hopes of staying secure behind the broad oceans, such attempts at geostrategic distancing did not provide effective insulation any more than they had in the Napoleonic Wars or World War I.

Waves stirred up by that rising storm began to wash ashore in January 1940, well before the fall of France, and brought the IPF concept back into the spotlight. Britain's ambassador to the United States, Lord Lothian, spoke to the Chicago Council on Foreign Relations, at the epicenter of American isolationism and Anglophobia, raising the issue of American involvement in collective security and international military organization. He defined sea and air forces as forms of police power, insisting "they must be in the hands of the democracies, and not of one power." Lothian elaborated on that theme at the year's end when he faced the epicenter of American isolation while addressing the American Farm Labor Federation.[13] In late June, the Wehrmacht's rout of the British Army and conquest of France further underscored the irrelevance of the League of Nations and added urgency to the search for a new international security system in both the United States and in Britain.[14] A literal torrent of plans from every quarter began and flowed unceasingly for the next four years, including many designs for associations of democratic nations along the lines of the United World Federalist concept, as well as new versions of a revived League of Nations with "teeth."[15] That flurry of visionary idealism began in the midst of the presidential election campaign of 1940, as Franklin Roosevelt sought an unprecedented third term. As the GOP split on the issue of isolationism, FDR brought some prominent Republicans into his cabinet, but he conformed, as he usually did, to public opinion polls in his neutralist rhetoric vowing again and again to keep America out of war. At the same time, he covertly provided Britain with military aid and approved the forging of links between America's fledgling intelligence apparatus and Britain's, and between the Anglo-American scientific communities. In August, Congress approved a fivefold increase in the Army's strength, including its air forces, and sanctioned the first peacetime conscription in American history. A month later, FDR approved the transfer of fifty obsolescent destroyers to the Royal Navy in exchange for bases in British colonies in the Western Hemisphere.

From that point until Pearl Harbor, American public opinion polls reflected a desire to stay out of the war while supporting the Allies against Hitler. Both the Republican and Democratic parties were divided on the issues of neutrality and intervention, and as the apposite "America First" movement and "Committee to Defend America by Aiding the Allies" gained momentum, discussions of postwar peace enforcement grew more serious and complicated. That was partly due to the fact that the American and British Left were divided from September 1939 to June 1941 by Stalin's nonaggression pact with Nazi Germany. Many followers of the Stalinist line in the United States opposed aiding Britain after France fell, while those of a more liberal and/or anti-Fascist bent broke ranks. For example, veteran pacifist Freda Kirchwey, editor of *The Nation*, waxed enthusiastic about creating a new international entity based primarily on American armed strength, in which "the combined forces of the Americas must be put at the disposal of the League."[16] Her views resonated with proposals for a postwar "international police power" in Europe[17] and for weakening national sovereignty to purchase "peace and security" under a "super state formed to enforce international law,"[18] although here, too, some critics deemed it a "dangerous" idea based on false analogies between the international system and the structure of societies within nations, or between nations and individuals, upon whom formal legalisms and constraints could be imposed.[19]

Early in the war, enthusiasm for collective security was faintly visible in initiatives emanating from organized religions, including a papal proposal of December 1939 and several initiatives by Protestant denominations closely affiliated with the peace movement.[20] While the IPF concept was not visible or very muted in early schemes, it began to appear in manifestos of various religious groups in Britain, the United States, and other nations. In general, those statements blended themes of social and economic equity and disarmament with a strong United Nations and a rejection of isolationism.[21] The IPF, long a theme in high school and college debates in the United States, was featured in primary education materials, as well as in the media, including radio serials and comic strips.[22] Although the latter were seen as superficial and trivial, comics had quickly become the most widely circulated American print media in the late 1930s. The IAF was a central implied theme in such comic strips as *Captain Midnight* and *Blackhawk,* which featured multinational bands of "superhero" aviators. Those paralleled the torrent of proposals for a successor to the League of Nations, many of which were based on a major alliance of democratic nations. The best known and most controversial of those was Clarence Streit's 1940 blueprint for an Anglo-American union, an uneven design that included both an IPF model and one based more on the French Foreign Legion as a prototype than other elaborate IAF concepts that appeared in Europe.[23] That did not sit well with Americans who had a dim view of Britain for various reasons.

As the presidential campaign intensified in the late summer of 1940, the plight of Britain, fighting alone against the Nazis and Fascists, came to the fore as the Luftwaffe bombed London and Nazi U-boats began attacking Britain's merchant ships in earnest. At the center of that vortex, the Royal Air Force's multinational elements comprised a kind of ipso facto IAF, although Noel Coward's ode calling on the British public to "lie in the dark and listen" to the RAF bomber streams ignored its ethnic diversity by referring to "English saplings with English roots."[24] The flight to Britain of many European countries' governments and fragments of some of their armed forces included thousands of European airmen, along with a great many from the Dominions, and some from the United States. Beyond those serving in RAF components like Bomber and Coastal Commands were sizable Polish and Czech contingents in Fighter Command, which played a key role during the aerial Battle of Britain, as well as the "Eagle Squadron" comprised of American volunteers.[25] That mingling continued, and two years later, in North Africa, the Desert Air Force included elements from the RAF, Royal Australian, New Zealand and South African Air Forces, Free French air units, and the Middle East Air Force of the U.S. Army Air Forces (USAAF). That led one observer, swept up in the spirit of the times, to describe those cosmopolitan aggregates as "the vanguard of the popular armies which will rise in Europe tomorrow."[26]

Despite the growing popularity of that vision and assurances from the "bomber barons" of the USAAF and RAF that they could bomb Germany into submission and, thus, eliminate the need to land Allied troops in northwest Europe, the air offensive did not go well. After shifting from day to night attacks in 1941, the RAF came under increasing pressure from many quarters to redeploy their bomber forces in support of armies and navies in the various theaters of war. Bearing down on its strategic bombing offensive led to no clear results and mounting losses. The airmen found bombing accuracy and navigation much thornier problems than they had expected before the war or were willing to admit publicly as things went awry, perplexed by much the same fuzziness of perception that bedeviled intelligence analysts in the Vietnam and Gulf wars and the bombing of Serbia in 1999. Individual British photo-interpreters assigned a wide range of values to photos of known levels of damage, and postwar British and American inspectors' estimates of damage done to German industry by the Allied air offensive varied up to 50 percent.[27]

As bomb damage assessment generated an especially stormy debate in British defense circles,[28] the tightening circle of that predicament contrasted with the growing enthusiasm in the Western democracies for air power serving as the mace of justice in the postwar world. While FDR favored regional defense agreements during his first two terms as president, before Pearl Harbor, he had shown some slight interest in a version of Teddy Roosevelt's Anglo-American international maritime police force but was cool toward the League of Nations.[29] Roosevelt was especially cautious in articulating his views on collective security after the political roasting he received for his 1937 "quarantine the aggressor" speech. In his Four Freedoms speech to Congress in January 1941, FDR left peace enforcement and international postwar structure well out of focus, as he did in August of that year while discussing postwar peacekeeping organization behind closed doors during his meeting with Churchill off the coast of Newfoundland. Although the communique of that conference, the Atlantic Charter, was neither a document nor a treaty, nor officially binding, it had substantial *gravitas*, nonetheless. As FDR's critics and isolationists complained, it was not an agenda for peace talks, like the Fourteen Points, but a list of joint war aims—even though the United States was neither at war nor an ally of Britain. Falling well short of the unambiguous declaration of war the British hoped for, the Charter was, nevertheless, highly provocative, since German and American naval forces were in combat in the Atlantic. During discussions that led to its issuance, FDR excised the phrase "effective international organization" in favor of a vaguer reference to the ultimate "establishment of a wider and more permanent system of general security."

After the United States entered World War II, Vice President Henry A. Wallace became the most ardent proponent of a postwar IPF, although Wendell Willkie, the GOP's 1940 presidential candidate, was not far behind. After losing to Roosevelt, the latter had become an ardent champion of internationalism and "globe trotter," and his account of global jaunts, *One World*, became a major best seller. Despite widespread reservations about what Clare Booth Luce deemed "globaloney," from 1940 to the end of World War II, more and more Americans came to favor Willkie's vision of enforcing peace with air power. Many Republicans and some Democrats remained isolationist, and Roosevelt remained characteristically vague on the question of postwar international order, maintaining maximum political sea-room on an issue that had produced bitter political infighting in 1919 and might well again. A peace treaty and postwar security arrangements would have to clear the U.S. Senate and Congress as a whole, respectively, and isolationism had not ended the morning after Pearl Harbor. Despite FDR's caution, some Cabinet members from both major parties along with other administration officials took a firmer internationalist stance. As the U.S. Navy battled German U-boats in the North Atlantic during the summer and autumn of 1941, Secretary of State Hull echoed Julian Huxley's proposal for a "Security Club" that would serve as the basis for forming an International Army,[30] while Secretary of the Navy Frank Knox offered a refurbished version of Theodore Roosevelt's vision of an Anglo-American international naval police force to suppress piracy and banditry.[31]

Such pronouncements, combined with America's apparent alliances with Britain, Russia, China, and roughly two dozen other nations after Pearl Harbor, raised apprehensions in some crannies of the American polity. The American South, a main bastion of the Democratic Party and racially segregated, had been more internationalist than other parts of the country between the wars. Those sentiments began to ebb, however, as new proposals for creating international police organizations looked less and less like Teddy Roosevelt's and Lord Lothian's visions of a "great racial federation, as of the Anglo-Saxon people," that is, an imperial naval police force. Many fresh models were based on global governmental structures whose "police" components would be multinational contingents of ground and air forces.[32] The possibility that such elements might be able to reach anywhere on the globe quickly and deploy multiracial forces had long been a source of concern to nationalists as well as segregationists,[33] and the latter, well entrenched in the U.S. Congress, opposed creating any kind of cosmopolitan force with broad police powers. Nor were their apprehensions eased by lyrical and liberal suggestions of a world army, organized on the basis of "interracial cooperation," which would back up local police in domestic disturbances, "safeguard elections," and "restrain our own countrymen abroad." Even more alarming was the assertion that "Only those . . . in the pits of ignorance" would "fail to see the validity of the movement partially to de-localize and de-nationalize order and justice."[34]

There was already some substance to that model. In World War II, the United States was linked under the United Nations rubric with such ethnically and racially diverse comrades in arms as the Nationalist Chinese, French Moroccan *goums*, the Brazilian Corps, the King's African Rifles, the Bengal Lancers, Mongols from Siberia, and Haile Selassie's Ethiopian forces. Idealized in films, posters, and the Allies' propaganda, that pattern was not popular in the American South, nor was the fact that many designs and models for the postwar world bore Russian, British, and Chinese hallmarks. While many Americans supported a solid postwar international structure, they did not see eye-to-eye on such details as forming links with the British and French Empires, the USSR, or

China, or giving smaller nations increased power. Some, like the distinguished historian Carl Becker, retained Theodore Roosevelt's strong sense of Anglo-Saxon camaraderie in asserting that "the preservation of the British Empire is, and since 1783, has always been a major national interest of the United States," but others, like noted journalist Dorothy Thompson, who favored "a common police force for the globe," were not inclined to "lift a finger to save the British Empire as presently constituted."[35] The underlying tension between the Western Allies, often smoothed over during the war, bubbled to the surface from time to time, as in a minor flurry when U.S. Marines raised the Union Jack over Tarawa, a British colonial possession, and in the louder outcries that followed the British jailing of Gandhi and many of his followers.

As noted earlier, most proposals for postwar collective security systems that appeared in Britain and the United States were new versions of the League of Nations, despite no solid agreement on why it had failed. The veteran political commentator Walter Lippmann, for example, saw Wilson's primary error as his failure to support regional alliances that would have evolved into a large, solid international order. On the basis of that conclusion, Lippmann proposed forming a "nuclear alliance" of the United States, the USSR, Britain, and China, from which an order of "liberty under law" would spread throughout the world. (He used "nuclear" as a synonym for core, since no atomic weapons existed at the time.) Lippmann was vague in dealing with the stickiest point—exactly how "all peoples" would be "compelled to observe" the "laws" of "the new order."[36] Henry Bamford Parkes' contrary suggestions were less elegant and concise and even more confused on the question of applying force. Assuming a new League of Nations would only be able to orchestrate international cooperation on issues of mutual self-interest but not have the authority to "coerce states which refuse to accept its decisions," he urged a "continued exercise of American power . . . to maintain peace and rebuild civilization throughout the rest of the world." Parkes preferred an Anglo-American alliance with shared defense costs and bases and "a navy and air force . . . each . . . not less than five times as strong as . . . any country . . . [except] the Soviet Union." That would allow Britain and the United States—he assumed Russia and China's implicit assent—to "immediately send their armed forces against any nation that commits an act of aggression," as unlikely as that seemed amid the carnage of World War II.[37]

After Pearl Harbor, the myriad designs for collective security systems were paralleled by more substantial steps in the higher echelons of the United States government. Two weeks after the Japanese attack, Secretary Hull ended the deliberations of the Group to Study the Organization of the Peace to form a more complex subagency, the Advisory Committee on Post-War Foreign Policy, also comprised of government officials and private citizens. When twenty-five countries at war with the Axis powers announced the Declaration of the United Nations on New Year's Day 1942, serious discussions included the possibility of forming a "Supreme War Council" along the lines of the structure created during the First World War, comprised of senior military officials from Britain, the United States, the USSR, Nationalist China, and "possibly the Netherlands." At the same time, a postwar IPF was a common theme in United Nations leaders' rhetoric and a salient feature in many designs for ordering the postwar world.[38] In the late spring, Hull took a further step by creating the Subcommittee on Security Problems, including State Department and military and naval members who would later play active roles at the 1944 Dumbarton Oaks Conference on United Nations organization. Its members covered much of the same territory explored in the Paris Peace Conference, in the League of Nations, and in the World Disarmament Conference 1932–1937, labeling the design for a contingent force "the French Plan."

Vice President Henry Wallace, in a speech to the Free World Association on May 8, 1942, injected a populist flavor into the issue of a formal postwar international order by proclaiming the twentieth century the "Century of the Common Man." That dovetailed with the rhetoric of other populists, like novelist John Steinbeck, who described the "bomber crew" as "really a bomber team . . . the kind of organization that Americans above all others are best capable of maintaining . . . truly a democratic organization."[39] Wallace's rubric also symbolically trumped the "American Century," a term coined a year earlier by Henry Luce, the Republican publisher of *Time, Life,* and *Fortune,* in an editorial in *Life* magazine. Nevertheless, three days after Wallace's speech, an article by an anonymous Royal Air Force wing commander in *Life* magazine set forth a detailed design for an IPF that would implement "Forced Goodness." While his blueprint resembled the designs of the 1930s, especially those of the New Commonwealth, it also included such radical proposals as ending military secrecy, carrying out random inspections, and rotating commanders of various nationalities. (Not surprisingly, English was to be its working language.)[40] Although the RAF officer's proposal generated little response, Wallace's speech did, in the forms of both fiery condemnation and fulsome praise. Alf Landon, the 1936 GOP standard bearer, for example, condemned it as a call for "civil war," while then-famous journalist Raymond Clapper deemed it the "Gettysburg Address" of World War II.[41] As radical as it seemed to many critics, Wallace's "Century of the Common Man" was in tune with Roosevelt's "Four Freedoms" presented in early 1941 (freedom from want and fear and freedom of speech and religion), as well as the Atlantic Charter and FDR's "Arsenal of Democracy" rubric. But it was also clear that the "Four Freedoms" were not working principles throughout the United Nations, certainly not in most of the Allies' imperial holdings, the Soviet Union, Nationalist China, nor in some Latin American nations. Such inconsistencies weighed little against what Walter Millis deemed "the martial spirit," and Wallace upped the rhetorical ante on May 8, when he described the war as "a people's revolution."

Although some Britons remained skeptical about creating an IAF,[42] on the same day Wallace spoke, British foreign minister Anthony Eden announced his government's support of Roosevelt's favorite model for postwar peace enforcement—"the four great powers" or "four policemen," which the president later presented to Soviet foreign minister Molotov during his visit to Washington, D.C., in the spring of 1942. FDR had defined it as something like "what we in the United States would call the sheriff's posse to break up the gang in" while insisting that "we are going to be rid of the outlaw this time."[43] The British Commonwealth, the USSR, the United States, and China would comprise "a peace system" that would assume "the main burden for the maintenance of peace and the main responsibility for the economic reconstruction of the world after the war."[44] The British Left soon played into that fugue when Herbert Morrison, Labour Secretary of Home Affairs in Churchill's coalition government, spelled out Eden's basic proposal in greater detail. Madame Chiang Kai-shek chimed in mid-June 1942 in a broadcast from Chungking, and two days later the Nationalist Chinese ambassador to the United States, in a speech at Madame Chiang's alma matter, Wellesley College, called for a "new world order" based on strong, international standing armed forces.[45]

In late June, differences over postwar international organization and policies were highlighted in the off-year congressional election campaign as Secretary of State Hull formed yet another study group, the Special Subcommittee on International Organization, charged with focusing efforts toward shaping, planning, and eliminating duplication. Many of its members would also be key players at the Dumbarton Oaks Conference on United Nations organization two years later. In a radio broadcast in late July, Hull urged, as he had before the war, the creation of "some international agency" which "by force, if necessary" could "keep the peace among the nations," even though such optimism seemed somewhat irrelevant if not naive in the wake of the long series of Allied defeats. A month later, amid the tumult of the especially hard-fought congressional election, a lengthy discussion between British foreign office and American state department functionaries produced the detailed Four Power Plan, which failed to stir Churchill's interest. He opposed including Nationalist China as one of Roosevelt's "four policemen" or considering it as any kind of major postwar power. While he lost that battle, the prime minister managed to secure first-tier status for France, despite his distrust of de Gaulle, and to keep smaller nations from gaining parity with the big nations. And despite their ongoing differences over regional security arrangements, Roosevelt and Churchill agreed in February 1943 at Casablanca that the "victorious powers" would "continue fully armed especially in the air, while imposing disarmament on the guilty," a view shared by other major Allied powers to varying degrees until after the war.[46]

Throughout 1942, Roosevelt remained vaguely supportive of the "Four Policeman" model while Vice President Wallace and other administration officials carried the "League with teeth" ball far down the field. Some Republicans joined in, and in a Labor Day speech, Sumner Welles, then Undersecretary of State and at the peak of his fame and influence with closer access to FDR than Hull,[47] responded positively to Republican leader Wendell Willkie's three-point Republican election pledge, which included a proposal for "some system of joint international force" that would include "an international police power . . . [a] United Nations" to "serve as the "nucleus of a world organization of the future."[48] Even though an internationalist consensus was crystallizing among both moderate Republicans and Democrats, Wallace's extreme rhetoric continued to raise hackles in some quarters. In November, for example, the vice president told a Soviet-American group that "willingness to support world organization to maintain world peace by justice implemented by force is fundamental to the democracy of the common man in these days of . . . the airplane, which makes it necessary to organize the world for peace, [but] also furnishes the means of maintaining peace." He offered a blueprint for a postwar IAF based on the assumption that "the United Nations" would "have such an overwhelming superiority in air power that we shall be able speedily to enforce any mandate whenever the United Nations may have arrived at a judgement based on international law"[49]—but he did not specify whom he meant by "we."

At that point, Roosevelt seemed to move closer to his more avid subordinates' views on postwar collective security structure by charging the Joint Chiefs of Staff to draft plans for an international police force, without concern for the constraints that might be imposed by national sovereignty. That enthusiasm soon faded.[50] Once again, FDR veered away from Wallace and growing public enthusiasm for a strong IPF based on air power, as he made broad and vague pronouncements on postwar international order. Welles became the principal ball-carrier for American involvement in a strong global postwar system, his more restrained demeanor and rhetoric, contrasting with Wallace's bombast, carried more weight among American business, political, and bureaucratic elites. As his star reached its zenith, Welles called for the United Nations to "undertake the maintenance of an international police power . . . after the war to insure freedom from fear to peace loving peoples" pending the establishment of the "permanent system of general security promised by the Atlantic Charter."[51]

That document set the stage for the initial formation of the United Nations five months later, at first only a loose aggregation of powers opposing the Axis. It grew in fits and starts as public enthusiasm increased for a strengthened postwar version of the League of Nations. In September 1942, for example, the British Liberal Party passed a resolution supporting the creation of "an armed international organization which must be strong enough to coerce even a great power" with the "means of taking coercive action promptly" and "the authority to enforce the settlement of international differences."[52] As the idea of a solidly founded UN gained momentum in the media and government circles, the IAF concept also grew in popularity. One advocate, arguing in Maoist terms that "political power is derived more from military force than from unsanctioned international agreements," posited that "a supernational air force under the control of a supreme central political authority is the ideal means by which to smother the fire of war at, or soon after, its outbreak."[53] It was the golden age of popularized geopolitics—a theme often featured by Henry Luce in *Time, Life,* and *Fortune* and tied in closely with the concept of the "Air Age" and polar "great circle" map projections. One detailed scheme based on a vision of "aeropolitics" proposed a "close-knit girdle of land bases for police aircraft . . . like so many scattered Switzerlands" with "five or six airborne armies of two or more thousand planes each, dotted around the world" that "nations would probably fear . . . more than the prospect of a general war which, without them, is almost certain."[54] Another such proposal designated an elaborate network of twenty strategic airports positioned to dominate thirteen key oceanic choke points, from which the "police power for . . . a United Nations system . . . an international aerial navy" would radiate.[55] Somewhat similar was a design based on "two distinct air police forces, one earmarked for duty in the European theater, the other for the Far East," with rotating command and staff assignments among "the qualified officers of the three or four air police units participating," and the "effective mixing together of British and American air staff officers" in the Mediterranean and in India and Burma.[56]

Gathering under the banner of opposition were such diverse foes as isolationist Republicans in Congress, moderates who preferred to leave preventing and dealing with aggression to regional alliances,[57] and Norman Thomas, the grand old man of American Socialism. He opposed any kind of world government or "world federation," distrusted Stalin and Chiang Kai-shek, and feared the rise of American imperialism. While not wholly opposed to a "World Police Force," Thomas was concerned that its members would be unable to transcend their national allegiances, citing Robert E. Lee's agonizing over which cause to serve in 1861.[58] His position was not too far in essence from that of South African statesman Jan Christiaan Smuts. Cautious as he had been in 1919 in supporting a weakly militarized League of Nations, Smuts now favored a strong American commitment to a postwar world organization limited to democratic states and based on mutual accord rather than a "military machine."[59]

Despite the optimism implicit in all that, harsh realities in the flow of military operations were casting dark shadows across what many saw as a brightening landscape. Although the Royal Air Force came closest to being an international air force in structure, and multinational ground and naval forces were deployed in all major theaters of war and on both sides,[60] there was no effective unity of command on the Allied side above the level of theaters of operations. In the later phases of the conflict, mounting frictions among the "Big Three"—Roosevelt, Churchill, and Stalin—made it clear that neither Europe nor any other part of the world would be "policed" by truly multinational standing forces after the war, all the designs of IPFs and IAFs notwithstanding. And the quality of those designs varied widely. FDR's vision of the "Four Big Policeman" seemed crude and vague in contrast with more coherent and well–conceived visions like Max Werner's NATOesque "continental organization in Europe" founded on "an international military police power" with "sole control of the air forces."[61] Yet the fuzziness of Roosevelt's concept was more attuned to the context of practical politics in which attempts to draw clear boundaries and definitions had often generated friction.

Despite the growing estrangement of the USSR and the Western Allies, popular enthusiasm for a strong peace enforcement apparatus continued to grow on the Right and Left in Britain and the United States. A surprising course change on the American Right came in 1942 when former president Herbert Hoover scrapped his design for a "Fortress America" and called for a neo–League of Nations with "teeth," including "a small air force [that] could impose the conditional peace and the subsequent programs." A year later, Hoover steered more sharply to port when he favored creating a permanent worldwide organization, which put him to the left of the new rising star of the State Department, Undersecretary Edward Stettinius, who prospered as Sumner Welles' fortunes declined and eventually replaced Cordell Hull.[62] On the far Left, Scott Nearing urged a world federation that would fill the vacuum soon to be created by the imminent disintegration of various powers' overseas empires[63] highlighting the paradox of the United States' fighting for a better world while remaining closely allied with two major imperial powers, two major dictatorships and several minor ones.

In another paradox, air power, often used to bolster imperialism between the world wars, was now sawing away at the fibers of that system, a trend also noted by geopoliticians.[64] Such contradictions put the framing of acceptable definitions out of reach,[65] as did the growing awareness that, as in 1918–1919, idealistic rhetoric was being swamped by power politics in anticipation of victory and washing away all the plans and hopes for a strong mechanism that would enforce international order and security.[66] In counterpoint to the surge in popular enthusiasm for the IPF/IAF concepts were muted debates within and among all of the "Big Three" powers' defense hierarchies and a more widespread backing away from the heartfelt resolutions made in the heat of battle and sting of defeat. At the end of 1942, Anglo-American war planners had agreed in general terms to "a highly sophisticated global defense" scheme, along with demarcation of intricate zones of control and coordination of civil airlines, but the U.S. Navy soon began to drag its sizable anchor despite fierce attacks from some air power enthusiasts.[67] The sailors were in an especially good position to influence policy shaping, since Roosevelt, an Assistant Secretary of the Navy in World War I, had named former Chief of Naval Operations Admiral William Leahy as his personal Chief of Staff. Nor were the sailors alone in that fight. Beyond obvious differences in style and in "roles and missions" between the Army and Navy, many generals as well as admirals had been skeptical or cautious about air power's utility. Paradoxically, Billy Mitchell's test bombing of warships in 1920 led Navy leaders to accept the potential of aircraft, albeit on a more pragmatic basis than the Army's airmen. Few naval aviators wanted separate air service, so there was little of the sharp infighting in the Fleet of the kind that wracked the Army as its airmen gained increasing degrees of autonomy with the creation of the Army Air Corps in 1926, the Army Air Forces in 1941, the Joint Chiefs of Staff in 1942, and finally, the U.S. Air Force in 1947.

While interservice rivalry has often been ascribed to the armed services' competition for lean funds in peacetime and for control of joint operations in war, their rivalry ran very deep in other respects, despite the ability of some soldiers and sailors to look beyond such parochialism. The service academies' attenuated rites of passage, symbolized by the ferocious rivalry surrounding Annapolis–West Point football games, produced a tendency toward short tempers, curt manners, and self-confident arrogance that generated friction in war and peace alike. Reflexive rivalry was publicly visible at several points during World War II, as when amphibious warfare increased the premium on cooperation in complex interservice operations. Some senior commanders in World War II like Generals Dwight Eisenhower and Omar Bradley, and Admirals Chester Nimitz and Raymond Spruance, appeared affable and low key. The most imperious were Douglas MacArthur and Mark Clark, while Army generals George Patton and Joseph Stilwell were "screamers."[68] Most irascible of all was Chief of Naval Operations Admiral Ernest J. King, who deemed himself an "S.O.B." In the Army Air Forces, some of the senior airmen were traditional curmudgeons,[69] but USAAF chief Henry H. "Hap" Arnold and senior American air commander in Europe Carl Spaatz, were more cheerful in public at least. Several young air leaders, like Hoyt Vandenberg, MacArthur's air chief George Kenney, and Philip Cochran, presented a very smooth, modernist demeanor. Youthful, athletic, and brash "flyboys" had been popular culture icons in the United States, and throughout much of the world since World War I, as living symbols of modernism. In the United States, those images were solidly fixed in the public mind by the media, and especially Hollywood, which relied heavily on aviation themes and symbols between the wars and during World War II. Not only were aircraft photogenic, but planes flying overhead were also the only visible evidence of the war for millions of Americans living away from the coasts, major camps, and maneuver grounds or not involved in defense production.

The roots of that symbol-making ran deep. Soon after World War I, Billy Mitchell launched a crusade for an independent air service, and the resulting schism in the Army officer corps was widened by his court-martial in 1926. Mitchell's principal protégée "Hap" Arnold, who rose to command the Army Air Forces in World War II, followed a different path than his mentor in appealing to public opinion. After leaving the service, until his death in 1936, Mitchell wrote appeals for enhanced air power aimed at general audiences, while Arnold and his colleagues studied journalism in night school and pitched a less strident message at the coming generation by writing books for young boys on the bright future of aviation. Despite their different approaches to publicity and some caution about Mitchell's intensity, his disciples were equally determined to reach their mentor's goal—a separate air force built around a core force of heavy bombers, which would win a quick, cheap victory by carrying out high-altitude, daylight precision bombing raids on key nodes in an enemy country at the outset of a war. Slow in coming, it seemed much closer at hand in 1935, when the Baker Board's recommendations led to the creation of the Air Force Headquarters (AFHQ), with a heavy bomber force under its control. That was much less than half a loaf, and as the fight for independence continued, the airmen in the Army faced opposition from many points on the compass, especially from traditionally minded ground forces officers. Showing little gratitude for receiving a relatively large share of the Army budget in lean times, the airmen built close links with the press and Hollywood and staged media events, including endurance and long distance flights and dropping water bombs on forest fires. At the same time, after becoming chief of the Army Air Corps in 1938, Arnold focused his noted personal charm on Roosevelt. During World War II, he would literally exhaust himself, and he died just after the war. While Arnold was selling air power to the public and trying to ease tensions between his airmen and the rest of the Army, most naval aviators saw themselves as sailors first. The Navy's upper echelons were not wholly resistant to the implications of aviation but did not fully grasp that aircraft had become a central element in waging war until after the Pacific Fleet's battleship force was shattered at Pearl Harbor.

During World War II, a sizable portion of the Army Air Forces assisted the Army's ground forces with cargo and troop transport planes, close air support fighter-bombers, and carpet bombing. They also supported the Navy in amphibious operations, reconnaissance, attacks on enemy shipping, bombing U-boat pens, long-range antisubmarine patrols, and mine-laying operations. Air power became the keystone of the U.S. Navy's force structure, even though almost half the Fleet was involved in conducting and supporting assault landings at the war's end, along with major transport, submarine, and antisubmarine operations. Beyond shore-based forces and air-sea rescue, the big task forces, built around the nearly hundred carriers of various sizes in service by 1945, were the superstars of the South and Central Pacific campaigns. Atop that vast complex of what would later be deemed "air assets" were the heavy bombers—B-17 *Flying Fortresses*, B-24 *Liberators*, and the very long-range B-29 *Superfortresses*—the mainstay of the USAAF's air war, in which strategy was entwined with a political campaign to gain independent status as an armed service. While many airmen clung to that model with the tenacity of religious zealots throughout the 1920s and 1930s, it did not become a main element in U.S. defense efforts until France's defeat in the summer of 1940 dramatically altered American strategic plans. Up to that point, U.S. airmen expected that the U.S. Army Air Corps would use French bases to launch an unmanned aerial vehicle—the "bug"—in large numbers against German cities. Throughout most of the 1930s, Congress opposed developing a true intercontinental bomber on the grounds that it was an aggressive weapon but, in 1938, funded the development of what became the Consolidated B–36 *Peacemaker*, a type of plane that did not enter service in 1949. After Pearl Harbor, air planners expected B-17s and B-24s to stand in the breach, flying from British bases until the much larger and longer-range Boeing B-29 *Superfortresses* came into service in 1943. President Roosevelt, however, diverted the B-29 to China as a gesture of support to Nationalist China's leader, Chiang Kai-shek. The B-29, bedeviled by serious engineering problems, was something of a stopgap itself, but its diversion placed the main burden of the strategic air campaign in Europe on the B-17s and B-24s of the 8th and 15th Air Forces.

During the early months of World War II, when it was clear that uneven Army-Navy coordination had contributed to the early defeats and disasters, there was a momentary lull in the infighting in Washington, D.C., and even less interservice wrangling throughout the war in the combat zones. Although Army Chief of Staff George Marshall denied his airmen a separate service, he offered vague assurances that it was likely to come after the war. Nevertheless, the Navy's hierarchy—King, Leahy, and Assistant Secretary of the Navy James V. Forrestal, who became "SECNAV" (Secretary of the Navy) after Frank Knox died in 1944—feared being outvoted if a third armed service was created. The sailors' anxieties rose sharply after Pearl Harbor, when Churchill and the top British military leaders arrived in Washington at Christmastime 1941 and prevailed upon Roosevelt to approve both a formal merger of the top British and American military command echelons into the Combined Chiefs of Staff and the creation of inter-Allied and joint service structures. At the same time, FDR approved the forming of the Joint Chiefs of Staff (JCS), which made King, Marshall, and Arnold virtual co-equals with their British counterparts. The arrangement was an anomaly, since the Chief of the Air Staff was the head of Royal Air Force, an independent armed service. Also, it further strengthened links between the Anglo-American airmen that had been forged before the war in a kind of pseudodiplomatic alliance. Finally, it altered the dynamics of U.S. defense planning, long based on a one-to-one Army and Navy interface that put a premium on negotiation and compromise. From 1919 to that point, joint doctrine and war planning, including the war contingency Color and Rainbow plans, had flowed upward for final shaping from the Army General Staff's War Plans Division and the Navy staff under the Chief of Naval Operations (CNO) to the Joint Board, which was comprised of senior officers from those agencies. There was, however, no formal American strategic planning structure, so that the services and, from time to time, the State Department, were engaged in filling an organizational vacuum. Although it provided some coordination, there was far less functional fusion and blending between U.S. forces than in Britain, or even Germany.

Early in World War II, after the JCS was formed, popular support for unification grew, presenting Admiral King, named CNO and Commander in Chief of the Navy immediately after Pearl Harbor, with several dilemmas. During the first six months of the war, the much reduced Pacific Fleet's battleship force was out of action, forcing the Navy to devise a new plan for waging war against Japan. The old one, War Plan Orange, and its successors were based on the expectation of a clash of surface fleets in the far Pacific. At the same time, German U-boats roamed American coastal waters virtually unchallenged for several months, while the Japanese overran Malaya, Thailand, the Philippines, Indonesia, and northern New Guinea, and their navy inflicted defeat after defeat on the American, British, Dutch, and Australian fleets. That torrent of disasters weakened the Navy's political defenses at home and marred its reputation far more than that of the Army or the Army Air

Forces, even though most of America's land-based air power in Hawaii and the Philippines had been lost during the first forty-eight hours of the Pacific War. As a result, in the six months following Pearl Harbor, as the Navy cobbled together a new war-fighting doctrine based on submarines, aircraft carriers, and amphibious warfare, the Army Air Forces gained some advantage from the sailors' predicament, as well as from President Roosevelt's call, before the United States entered the war, for a massive aircraft production program aimed at what many then felt was an impossible level—50,000 planes a year. (Ultimately, it reached over 90,000.) The USAAF gained further momentum from Pearl Harbor to the North African landings almost a year later, when the Anglo-Americans were nearly inert strategically. During that period, the potential hazards and logistical constraints involved in a major assault landing in northwest Europe by Allied ground forces, as well as inter-Allied wrangling, left the burden of bringing strategic force against Nazi Germany on the Royal Air Force's Bomber Command and, later, on U.S. Army Air Force heavy bombers based in England. Amphibious warfare in the Central Pacific did not begin in earnest until late 1943. Amid that inertia and frustration, calls for increased interservice cooperation increased, even from such naval luminaries as Admiral Chester Nimitz, commander of the Pacific Fleet, and Admiral "Bull" Halsey.

The unification squabble, like the fight over an independent air service, had long roots. Between the world wars, the Navy and its friends in Congress, most notably House member Carl Vinson, derailed a series of bills proposing unified armed forces. But reports of poor cooperation between the services contributing to the string of defeats and disasters during the era of "too little and too late"—December 1941 to November 1942—led junior Army and Navy officers in the newly built Pentagon to devise a highly visible "joint" operation, the "Doolittle Raid." As a result, in mid-April 1942 Army medium bombers launched from a Navy aircraft carrier made the first American air attack on Japan. After the battles of the Coral Sea and Midway in May–June 1942, a naval war of attrition began near the Solomon Islands, lasting from August until late 1943. From the beginning of the Pacific War, the triumphs of Japanese and American carrier forces supported the arguments of those who saw air power as the only immediate way to strike back for the foreseeable future—and the cheapest to boot. At the same time that the first American bomber units arrived in Britain in midspring 1942, public clamor mounted on both sides of the Atlantic in favor of a "Second Front Now"—a major Anglo–American landing in northwest Europe that would take the pressure off the beleaguered Red Army. At a lower volume were demands for a new unified defense organization that more properly reflected the growing importance of air power in waging war.

When a major War Department reorganization made the Army Air Forces one of its three main components, many observers saw that as a major move toward eventual autonomy. USAAF chief Arnold would struggle throughout the war to make sure Marshall's implicit assurance was not reversed. Well before nuclear fission proved feasible, Arnold covered a spread of bets by putting his personal weight behind creating a special force to deliver atomic bombs and by pressing home the strategic bombing offensives against German and Japan.

The assumption of winning the final victory was implicit in most of the IAF and IPF plans that appeared in the media in the United States and Britain from 1941 on, as well as in official circles. That was not only remarkable during the early, dark days but also later on since, despite optimistic rhetoric and the Casablanca "unconditional surrender" proclamation, an all-out Allied triumph was not absolutely certain until relatively late in the World War II.[70] Not surprisingly, amid all the uncertainty and apparent stagnation in 1942, some saw visions of a viable form of international government preceding "compulsory arbitration, collective security, and an international political force" as optimistic and irrelevant.[71] And yet, many Allied policymakers disagreed. Some preferred a postwar international system along the lines of Alexander Hamilton's view of the Federal Union, based on mutual interest and consensus rather than coercion.[72] However, rising public enthusiasm for a model with "teeth" reflected the deep resolve and optimism of most Americans that the Allies were going to fight on through to what Roosevelt had called "the inevitable victory, so help us God," and then build a solid foundation for maintaining and, if need be, enforcing peace by using military force.

The spring of 1944 marked the apex of the IPF and IAF concepts in World War II and, indeed, in the twentieth century. Both Wallace and Welles were about to pass into political oblivion, the former dumped in the summer by the Democratic Party's moderate and conservative politicos, while the latter was edged out of the State Department. Beyond a fierce personal animus between Hull and Welles, they differed strongly on the structure of postwar international order. The secretary of state preferred a large, worldwide system; Welles, a smaller regionalist model. For roughly a year, the latter had been riding high, widely viewed as a "fair-haired boy" and a favorite of Roosevelt, who also seemed to prefer regionalism. Welles' writings were widely read, many journalists supported his views, and even though Hull shunted him out of a key planning role, the first draft of the UN Charter was based on Welles' regionalist model. Ultimately, the Secretary of State was able to sack his rival on the grounds that he was gay. Amid that squabbling, FDR momentarily expressed support for a global postwar order enforced by a "universal peacekeeping system,"[73] but before Wallace and Welles left the high battlements, both had moved much closer to the president's vagueness in making statements on postwar collective security."[74]

Despite infighting and ambivalence in the corridors of power, the internationalist bandwagon gained speed throughout the war. The issue of the postwar international order was dramatized in the spring of 1944, when General George S. Patton Jr., awaiting the Normandy invasion in England, got into serious trouble for endorsing a postwar Anglo-American global peacekeeping system.[75] Despite his chronic rashness, Patton's confusion regarding the matter was understandable. Beyond Hull's comments early in the war, Roosevelt had spoken in favor of "an international police force composed of the United States and Great Britain" as a step toward creating a broader world organization, and the United Nations was visibly evolving into a more solid structure when Patton offered his views to British ladies at a garden club.[76] Despite FDR's support for the "four policemen" model, Churchill and many others in Britain and the United States had clung to a vision of Anglo-American fusion in the spirit of the Prime Minister's dictum that: "If we are together, nothing is impossible. If we are divided, all will fail." The creation of an IPF and an IAF seemed imminent as serious drafting of the postwar international system's organization chart at Dumbarton Oaks in August 1944 drew closer.

Beyond those disparate views, the IPF and IAF concepts dovetailed nicely with strategic bombing advocates' visions and goals. In mid-1943, General Marshall had ordered War Department military staffs, including the Air Staff Plans Division, to work on postwar plans for American participation in "an international military force," under the assumption that maintaining a sizable contingent of U.S. ground forces was not in the cards.[77] In the political domain, public opinion polls showed support for an independent air service, and an international air force. As Moss Hart, co-producer of *Winged Victory*, and journalists like Al Williams and Alexander P. de Seversky beat that drum loudly, filmmaker Walt Disney had joined in at a much higher volume. Putting his bread-and-butter training film contracts with the Navy at risk, he produced an animated film based on de Seversky's best-selling polemic *Victory Through Air Power*. The book assigned a constabulary role to the Army, and made the Navy into a super Coast Guard, although Disney's film left that aspect out of view. At the same time, Ely Culbertson, a well-known tournament bridge champion, presented a detailed grand design for an IPF and IAF in a book that became a best seller. Rejecting both neo–Wilsonian models and Clarence Streit's regionalism, he offered an elaborate World Federation plan much like many proposals presented at Geneva by various nations between the world wars. It included levies of men and money on nations for a two million strong "world police," basing schemes, and detailed rules for arms control and recruiting.[78] Congress chimed in on a fainter note by creating a subcommittee of the House Committee on Naval Affairs to study postwar Pacific basing.

Enthusiasm for such models continued to grow as the war approached what Churchill later called "the hinge of fate" in mid-1943. Although British and American casualties and costs were rising sharply, the panicky sense of "too little and too late" and fears of Axis triumph that haunted the Allies earlier were fading. On the other hand, at that point (two-thirds of the way through the war as it turned out), the Nazis still dominated most of Europe and had advanced to the Caucasus and Volga, while tightening their grip on Leningrad as Admiral Doenitz's U-boat wolfpacks were shredding Allied convoys in the Atlantic. Italy was still in the fight, and Japan controlled much of the western and southern Pacific, as well as eastern China and Southeast Asia, while threatening India and parrying Allied advances in New Guinea and the Solomon Islands area. The only real bright spots at that moment were the Red Army's victory at Stalingrad, the Western Allies' capture of Tunisia, and the first signs of a turning tide in the Battle of the Atlantic. The U.S. Marines had not yet made their first major assault landing in the Pacific, and highly publicized efforts like the Doolittle raid in April, American bomber raids on France from British bases, and the Marines' landing on Guadalcanal in August were very small in scale. In the Pacific, the Japanese still menaced Australia, despite setbacks in the Coral Sea and at Midway, and had pushed through Burma to India's frontier while MacArthur's Australian and American forces clung to a foothold in New Guinea. The first major Anglo-American counterstrokes had come in late October and early November, as Montgomery's victory at El Alamein in Egypt came just before the major Allied landings in Northwest Africa—Operation TORCH. But there was still no second front in western Europe, despite loud demands for an invasion from the Soviets and many in the United States and Britain since the late summer of 1941. In spite of high hopes, the Anglo-American bomber offensive against Germany was about to run into serious trouble.

Nor were things all that bright in the political realm. Throughout 1942, both Churchill and Roosevelt had been under very heavy rhetorical fire, as the Allies seemed hopelessly mired in organizational confusion and inertia and were rarely able to win a battle. Some Republicans wanted to suspend FDR's constitutional role as commander in chief of the armed forces and bring General MacArthur back from Australia to oversee the U.S. war effort as a generalissimo. Others proposed forming a war cabinet like Britain's or the *Oberkommando des Wehrmacht*, hoping to rein in Roosevelt's free-wheeling and vague style. On the surface, sizable GOP gains in both houses of Congress in the congressional elections—the day before the TORCH landings in North Africa—seemed to reflect widespread discontent with the Roosevelt administration's management of the war effort. However, turnout was low, in good part due to the fact that literally millions of Americans were in military service or had moved to take war production jobs and had no recourse to absentee ballots. In any event, since almost all isolationists seeking reelection in 1942 retained their seats,[79] an expedient coalition of Republicans and antiliberal Democrats held sway in Congress for the next two years. At the same time, some conservative Democrats were internationalist on some issues, and public enthusiasm was growing in favor of creating an IPF and America's playing a major role in maintaining world order after the war. That trend became boldly apparent in an Associated Press poll in the spring of 1943, which showed over 75 percent favoring a "League with teeth," in contrast with a 23 to 32 split among GOP senators against creating such an organization.[80]

The hinge soon turned, politically and militarily, in midyear when the Allies won further major victories—the Soviets' triumphs at Stalingrad and Kursk, the Western Allies won a clear victory over Nazi U-boats in the Atlantic, and the Anglo-American Combined Bomber Offensive against Germany began in earnest with high expectations of success. The latter stemmed from Churchill's tense meeting with Stalin in Moscow in August 1942, during which the prime minister promised intensified both RAF bombing raids against Germany and the TORCH landing in lieu of a second front in 1942. At their Casablanca meeting in February 1943, Roosevelt and Churchill gave that assurance major substance when they approved a plan for a major "round-the-clock" bombing offensive based on British night raids and American daylight attacks. The prime minister's thunderous rhetoric echoed that shift in strategy when he spoke before the American Congress in mid-May. Painting a grim picture of the consequences of the impending air offensive, he warned that "if the munitions populations [of Germany] . . . do not like what is coming to them, let them disperse. . . . The process will continue ceaselessly with ever-increasing weight and intensity until the German and Italian peoples abandon or destroy the monstrous tyrannies which they have incubated and reared within their midst."[81] That speech was met by as loud a round of applause as his proclamation a month later at the Guildhall in London: "that superb but awful force [Bomber Command] . . . will be applied to the guilty nation and its wicked leaders. . . . Never was there such a case of the biter bitten."[82]

Major leaders on both sides commonly spoke in ferocious terms, but on the Allied side, such gleefully murderous and vengeful rhetoric contrasted sharply with idealistic visions of the postwar world order. Like a major storm and as in other major modern wars, the longer World War II raged, the nastier the polemics became, careening along their own course and gaining a tempo that confounded predictions, often dashing the hopes of those who tried to control or brake it. That was as true on the home front as it was both in the combat zones and in the highest echelons of power. When Allied triumphs and the approach of the Dumbarton Oaks United Nations Organization Planning Conference led Secretary of State Hull to suspend the deliberations of various postwar planning committees in the summer of 1943,[83] a midsummer poll in the United States mirrored those in Britain showing that three-quarters of those queried supported an international police force to maintain the peace.[84] In contrast with that trend, however, Vice President Wallace eased off his earlier position by suggesting that "the air arm of the United Nations peace force" should not be a large standing force. Pointing to high costs and long intervals of disuse, he proposed that it be comprised of military airmen, who would serve after initial training "as the air and ground crews of the United Nations air network, which in peacetime would be commercial and carry passengers and freight." Although he conceded such a structure would require American subsidy and the building of many fighters and bombers, Wallace argued that it would be cheaper and more useful than "a military air force in idleness."[85] On the other hand, Churchill, although a staunch Tory on some issues, remained committed to an IPF, in keeping with both his willingness to shave national sovereignty and his support of an international air police force since 1919, especially as president of the New Commonwealth in mid-1930s.[86]

Beyond the fact that all IAF and IPF designs that appeared in books, media, and speeches at that time ended up the scrap heap of history, many were obviously grandiose, idealistic, or simplistic. Nor was there a consensus among journalists. For example, Raymond Gram Swing, a popular news commentator, favored "enforcement of law on a neighborhood basis," while on a more cautious and less folksy note, Henry Wriston warned that relying on force to keep the peace not only put undue faith on an irrational modality but also threatened to create a "police mentality" and "habits of thought which are the antithesis of those appropriate to either freedom or democracy."[87] Those demurrals had little effect on the growing sense of resolve among the general public in the United States and Britain, born of the early defeats and mounting carnage of the war and a widespread feeling that it might have been somehow averted, that something had to be done to put it right. That attitude, increasingly visible throughout the domain of popular culture, was articulated by Dorothy Thompson and other intellectuals in a spirit similar in essence to Irving Berlin's song "This Time Will Be the Last Time."[88] But the enthusiasts were not in step. Thompson deemed Culbertson's blueprints "deceptive, unrealistic, and reactionary," and a "nice mess" that resembled the failed, limited visions of 1919, and she called for "a mutual association of nations."[89] And some opposed all forms of collective security. An international superstate was anathema to such bulwarks of American patriotism as the National Association of Manufacturers and the American Legion, as well as isolationist Senators Robert Taft and Hiram Johnson.[90] As the war turned in favor of the Allies, however, public enthusiasm for Wallace's and Willkie's visions increased, producing a rift among Republican House members, 40 to 115, in favor of a vague "Postwar Manifesto."

In April 1943, the changing mood in America's heartland was reflected in Senator Joseph Ball of Minnesota's proposal for "some limited form of government on a world level." Arguing that since "control of war demands strong international action," Ball suggested an "air force should form the backbone of the international force"[91] as he co-sponsored the Ball-Burton-Hatch-Hill Resolution (B2H2), which called for the United States to lead the way in creating a formal United Nations structure.[92] Although it was not submitted for a vote in committee or on the floor, B2H2 focused public attention on the issue of using force to maintain peace in the postwar world. So did a resolution offered by Democratic congressman J. William Fulbright of Arkansas in September 1943, approved by the House, 360 to 29, which called for the "creation of appropriate international machinery with power adequate to establish and to maintain a just and lasting peace . . . through its constitutional process."[93] Those proposals, forged when memories of "too little and too late" and the long series of very close calls were still fresh, left the thorny question of sovereignty out of focus.[94] That omission led a scientist to warn that "if we subject any major portion of the world to the control of an International Police Force, we, ourselves, should accept a similar control on the same general principles."[95]

In June 1943, James Shotwell and Clark Eichelberger, members of the Group to Study the Organization of the Peace formed by Hull in 1939, were named to oversee a team that worked on a United Nations Charter draft for presentation at the Allied Foreign Ministers Conference in November. Accepting sovereignty as a given, they included a proposal for "an international air police force recruited by voluntary enlistment from various countries" under UN auspices.[96] As Congress began to change course in an internationalist direction, concern for the Senate's constitutional treaty ratification role and national sovereignty led Texas Democratic senator Tom Connally to offer a resolution much like Fulbright's, calling for U.S. cooperation with other nations to establish "an international authority with power to prevent aggression and preserve the peace of the world . . . a general international organization, based on the principle of the sovereign equivalency of all-peace loving states, for the maintenance of international peace and security." Based on "Big Three" joint official statements, it restated the constitutional requirement that U.S. involvement in a world organization would require a two-thirds majority vote and gained Senate approval, eighty-five to five, with five absent.

However, when Roosevelt offered his "four policemen" concept to Stalin in November 1943 at the "Big Three" conference in Teheran, the Soviet dictator, as the head of a vast Eurasian empire, shared some of Churchill's misgivings about allowing China too large a role in world affairs.[97]

The postwar international system was also being discussed earnestly at the grass roots, from explanations to secondary school students of the complex issues involved in *Scholastic* magazine,[98] to further schemes devised by scholars and journalists.[99] Out of the public's view, "devils" in the details of blueprints for an international organization came into sharper focus in official circles as Allied diplomats and military leaders began grappling with such thorny problems as occupation, joint policing, and arms control. At Teheran, for instance, Roosevelt assured Stalin that he wanted American armies out of Europe after the war but reserved the option to maintain a substantial American air force overseas.[100] Over the next two years, however, the president meandered on the issue of postwar order, much as he had during the drafting of the Atlantic Charter, even though as a member of the "Big Three" and "Big Four," he routinely issued joint statements, like the Moscow Declaration of November 1, 1943, which seemed to commit the United States to join Britain and the USSR in "joint action on behalf of the community of nations." But like the Atlantic Charter, that declaration was not a treaty, nor were the State Department's Commission to Study the Organization of Peace's detailed reports, published throughout the war, which included a proposal for an international "mobile policing force under the command of authorities responsive to the world community as a whole." Similar intentions were visible in a vast array of studies, public and private, official and unofficial, open and secret, that addressed various problems of using force against aggression,[101] offering models, including standing units, and "national contingents" under an international "Combined General Staff" which would prepare war plans, and oversee the ad hoc mustering of forces.[102] That model became increasingly popular, since it required no clear multinational structure, no international funding, nor the defining of complex ground rules required to form and deploy a standing force, whatever its details. Beyond that, it assuaged fears of internationalist Bonapartism.[103]

The increasing flow of proposals for creating IPFs and/or IAFs, however, did not clarify matters. By late 1943, it was not only less and less certain how an IAF would be used to enforce peace but also increasingly unclear whether that was possible at all. The Spanish Civil War, China, and the Battle of Britain had made it clear that strategic bombing was not a war-waging panacea, but another form of attrition warfare. The Anglo-American "bomber barons' " hopes had been dashed as their respective air offensives in Europe were fought to a momentary standstill by the Luftwaffe at different points during 1943.[104] Their chance to deliver the victory without the opening of the Second Front by bombing Germany had passed. With both the Western Allied ground forces in Italy and the Red Army advancing, and the Normandy invasion imminent, strategic air power was no longer the main means of attacking the Third Reich. Later, British official historians summed up the unhappy state of affairs thus: "the limitations of the force were only gradually discovered, and for more than two years Bomber Command, in spite of a few remarkable successes, was to a great extent lost in the dark, the haze and the searchlight glare."[105]

Things were not going much better in the daylight. One basic problem was that both sides were continually introducing technical innovations that canceled each others' advantage. Another was that Albert Speer, Hitler's Minister of Armaments, had reformed German war production, yielding much greater numbers of fighters and better antiaircraft systems. As British and German jet fighters were coming into production, albeit in relatively small numbers, the Anglo-Americans were producing increasing numbers of long-range escort fighters and bombers and radar bombing aids, but those were slow in arriving, forcing the 8th Air Force to stand down in October 1943. Despite all the setbacks, the Allies' air attacks damaged German industry extensively and ravaged its cities, forcing the Nazis to divert more and more of their resources to air defense. About 5 percent of *Reichspropagandaminister* Goebbels' diaries dealt with the Allied air offensive, which he saw as "our most vulnerable point."[106] But as deep as those wounds were, the three-month halt in American daylight operations and a pause in Royal Air Force Bomber Command raids earlier in the year were increasingly bigger beads on the lengthening string of the failure of large-scale bombing operations to produce decisive results.

That gloomy course of events also bore out the prediction of Admiral Thomas Hart, commander of the ill-fated U.S. Asiatic Fleet early in World War II. In early 1943, while chairman of the U.S. Navy's General Board, Hart advised the Secretary of the Navy that creating an effective IPF was years away. Soon afterward, the U.S. Navy's hierarchy launched a political counteroffensive in the teeth of growing public enthusiasm for establishing an air force separate from the traditional armed services and an international air force after the war, and for unifying the armed services under some kind of national defense agency or department. In the summer, Admiral James O. Richardson, Pacific Fleet commander before the war, headed a committee to study postwar unification and signed the minority report opposing the concept.[107] Even though the Combined Chiefs loosely agreed that some sort of international peace-keeping forces should be created after the war and authorized detailed joint planning for its basing system,[108] the Navy was able to keep Army Air Force planners out of the planning process until 1945, after the design of United Nations military components was completed.[109]

However, the sailors' tactical successes in the corridors of power did not nullify the effects of public opinion, nor did the setbacks in the air campaigns against Germany and Japan in 1943–1944 dull the American public's confidence in strategic air power or its enthusiasm for collective security. But the will of the masses weighed no more in the shaping of policy when the Dumbarton Oaks United Nations organization conference began in Washington, D.C., in late summer 1944 than it had in 1919, when Woodrow Wilson arrived in Paris. Nevertheless as the conclave convened amid the furor of Roosevelt's precedent-shattering attempt to seek a fourth term as president, international organization and military mechanisms for keeping the peace were central items on the agenda, in apparent congruence with public sentiment, and imminent major political sea changes.

That was visible in the realm of Democratic Party politics. The principle behind FDR's invoking his role as commander in chief in the election slogan "Don't Change Horses in Midstream" was not extended to his vice president. Widely viewed as too liberal, Wallace was shoved offstage, and Senator Harry Truman of Missouri was named as the Democratic vice-presidential candidate. Despite that, many ardent liberals continued to support Roosevelt. In Hollywood, stars like Orson Welles took the stump for FDR, while a highly touted but signally dull and commercially unsuccessful film, *Wilson*, portrayed the Senate's rejection of the Covenant of the League as paving the way to World War II. At the same time, polls showed strong public internationalist sentiments, and even some former isolationists, like Senator Arthur Vandenberg, now supported strong collective security measures.[110] Nor did candidates debate the IPF–IAF issue, aside from some marginal skirmishing. The Republican presidential candidate, Thomas Dewey, an Eastern moderate internationalist, steered much closer to Willkie's and the Roosevelt Administration's course than his main rival for the nomination, Robert Taft of Ohio. The latter, deemed "Mr. Republican" and leader of the Party's conservative wing, veered sharply away from the course set by his Progressive Republican father, who had headed the League to Enforce Peace and championed an international police force during the First World War. The younger Taft opposed the B2H2 resolution, as well as globalism in general, and managed to insert favorable references to sovereignty into the Republicans' Mackinac Island policy declaration, even though he went along with its statement of general support for a United Nations organization. But his lack of enthusiasm for that put him well out of step with the public mood. While favoring "covenants outlawing aggression," Taft firmly opposed the United States making firm commitments to enforce them or deferring to any "President Whoozis of Worlditania."[111] In early 1944, during the approach to the Republican convention, his resistance sounded lonely and futile amidst the mounting clamor for an IPF and/or IAF and the neo-Wilsonian themes appearing in the media.[112]

At the grass roots, reviving the League of Nations had become a major theme in high school and college debates,[113] and the terms "world order" and "new world order" were scattered throughout yet another wave of treatises. Like their predecessors, those ranged from relatively serious-minded and well-conceived essays to vague, lyrical, and fanciful musings that limited the architecture of peace organization only sketchily.[114] An American Historical Association pamphlet prepared for U.S. Army troop education discussions in 1944 struck a balance in a long and balanced discussion of the pros and cons of raising an IPF, including the dilemmas of which laws would be enforced—and how.[115] The views of historians, however, diverged widely. Carl Becker called for a League "with teeth"[116] but, as Charles and Mary Beard pointed out soon afterward, even though "The number of Americans who thought that some kind of alliance or world association would be formed to keep the peace was very large . . . no simple or intricate plan that was set forth commanded wholesale approval."[117] On contrary tacks to the Beards' caution, Harry Lewis Braham attributed the League of Nation's failure to its lack of a "police force or power to compel the allegiance of the signatories" and suggested "a Union of Nations with one central army,"[118] while Dexter Perkins judged "the fundamental weakness of the Treaty of Versailles" as "the absence of an effective organized force for the preservation of the peace that had been resorted."[119]

Objections to proposals for a strong international peace enforcement mechanism were much like those heard a generation earlier—it would threaten sovereignty, lead to the emergence of a military superpower, allow some nations to manipulate the use of its police power against others, maintain the status quo through force, or perpetuate imperialism.[120] Others raised constitutional constraints or addressed intricate and often contradictory aspects like liability,[121] legitimacy, and difficulties presented by semantics and definitions, including "the unbridgeable chasm between *alliance* and *federation*."[122] Opponents also identified such mechanistic obstacles as difficulties in matching means to ends and fine-tuning the application of force,[123] the political constraints—domestic, international, and regional—of committing forces in remote and unpleasant settings, and problems in selecting specific coercive methods.[124] Perhaps the most tireless proponent of international federation, Count Coudenhove-Kalergi rejected the idea of an IAF, per se, proposing instead a system of regional alliances focused on disarmament, free trade, and improved health and educational standards, under the shield of an "American supremacy of the skies" analogous to the Royal Navy's role in the *Pax Britannica.*[125] Less lyrical but similar in essence was Henri Bonnet's plan for an IAF or "any other suitable" name "for air forces under a common authority" linked to an international and regional systems of civil aviation.[126] Both Coudenhove-Kalergi's and Bonnet's views aligned with those of Secretary of State Stettinius, Walter Lippmann, and increasing numbers in the American establishment.

Many such aspects of the postwar world order had been addressed in major inter-Allied conferences prior to Dumbarton Oaks. In July 1943, Secretary Hull had erected yet another peace planning edifice in the State Department's peace planning structure in January 1944—the Office of Special Political Affairs—which, like its predecessor organizations, was enthusiastic about creating an IPF and an IAF. So was Churchill. On the eve of the Normandy landings in June 1944, with Dumbarton Oaks two months away, he advised the House of Commons of his government's intention of setting "up a world order and organization, equipped with all the necessary attributes of power, in order to prevent the breaking out of future wars or the long range planning of them in advance . . . a World Council . . . [and] a certain minimum standard of armaments . . . ample Armies, Fleets and Air Forces available to prevent anything like that."[127] His tone meshed with the hopeful spirit running through the media, servicemen's letters and diaries, and slogans like Wallace's "Century of the Common Man," Willkie's "One World," and Henry Luce's "American Century." Lord Beveridge, architect of Britain's postwar welfare state, who shared Churchill's enthusiasm for an international peacekeeping framework, made perhaps the first reference to the United States, Britain, and the USSR as "super-powers."[128] (China and France were allowed into the inner circle only at the sufferance of the United States and Britain, respectively.) Like Roosevelt, Beveridge rejected the IPF model, describing peace-keeping mechanisms in grandly vague terms as a "supra-national agency with preponderant force at its command" under the authority of a "self-constituted coalition of first-rank powers as the nucleus of a general security program."[129] While American military leaders, aside from Patton, usually refrained from commenting on such issues, some senior British airmen echoed Churchill, including Wing Commander Guy Gibson, hero of the "Dam Busters" raid, and Air Vice Marshal Don Bennett, commander of the RAF's Pathfinder Force, whose treatise *Freedom from War* called for creating an IAF.[130] Canada's top World War I ace, Air Marshal William Bishop, matched their intensity in urging "something close to world government" and an aerial police force to oversee the internationalization of air power and commercial aviation, declaring that he "would not waste one moment in talking before raising plain, ordinary hell with anybody in the world who committed any infraction of the rules of peace, in fact or in spirit."[131] Bishop's model, like many others, reflected the interwar fears that commercial air services could mask strategic intentions and be a threat to peace in his proposal of internationalizing airlines as a solution.[132]

Divergences on the IPF and IAF issue widened as imminent victory led the major Allied powers to look more closely at the practical aspects of what their national interests would be in the postwar world. The growing strength of America and the Soviet Union shaped both rhetoric and policy, producing a mix of double standards. For instance, the intensity of condemnations of imperialism was inversely proportional to the size of the great powers. Anti-imperialism became a major theme in American journalism and diplomacy during World War II, while British and American leaders, despite the relative closeness of their wartime alliance, clashed constantly over wartime strategy and postwar spheres of influence. The Soviets' rhetorical sympathy for downtrodden peoples was not extended to those under their suzerainty, and as in 1919, special American interests like colonies, the Monroe Doctrine, and the Roosevelt Corollary were left out of focus.[133] In the course of the war, the United States had wrung accommodations and treaties from Britain, beginning with the pre–Lend Lease sell-off requirement that set the stage for postwar American dominance in such areas as aircraft production, international air commerce, and motion picture distribution. As the war progressed, an ever harsher spotlight was aimed at the British and French empires, although access to imperial basing and communication networks provided a major strategic advantage to the Allies in fighting a global war, and American forces assumed a surrogate role as colonial overlords in lieu of British forces in the Caribbean. Those divergences fed the squabbles among the major Allied powers over structure at the Dumbarton Oaks Conference that peaked in discussions over UN trusteeships. France was given a role as one of the postwar "Big Five" powers on the United Nations Security Council and pressed once again for a strong international security structure including a general staff but not a standing multinational force *a la* the Foreign Legions of France and Spain.

Beyond those international differences lay the clashes between the American armed services that shaped the flow of events at Dumbarton Oaks. The U.S. Navy's hostility toward an IAF extended to the idea of an Anglo-American naval IPF as well, even though that concept was popular early in the war.[134] Deliberately or not, FDR had legitimated the British Empire in 1940, when he gave the Royal Navy 50 obsolescent destroyers in exchange for 99-year leases on bases in British colonies in the western hemisphere. In 1942, Nicholas Murray Butler cited the House of Representatives' Joint Resolution of 1910 as a precedent calling for the creation of "combined navies" or other collective means to keep the peace.[135] Another major shoal in that channel was the blunt Anglophobia of Admiral King and many other senior American officers, naval and otherwise, even though some American advocates of an IAF saw British "air policing" as a viable model for postwar peacekeeping and built conceptual models around the assumption of substantial Anglo–American power sharing.[136] Remnants of Teddy Roosevelt's naval international police force model were also visible in popular culture models of postwar international security. As the Dumbarton Oaks Conference convened, for example, a feature article on the postwar military balance in *Popular Science* identified "one realistic fact" standing "apart from diplomatic finesse and hopeful dreams of international organization"—since the U.S. and British fleets would rank first and second respectively: "What could be more natural . . . than to co-operate in policing the seas?"[137] The vision of Anglo-American world policing was also threaded through the final testament of Lord Davies, the dean of British hardheaded pacifists, who died in 1944. His detailed plans were based on a "Temple of Peace" supported by "seven pillars of peace," the pillar of force being based on wielding of air power.[138]

Other British proposals for revitalizing the League of Nations looked much like those devised by their American "cousins" and reflected many similar paradoxes. The tireless pacifist Norman Angell, for example, called for a new and tougher League of Nations, arguing that "unless the [world] community is prepared to use its combined power for the defense of the individual member who is made the victim of lawless violence, there can be neither law, nor peace, nor justice, nor stable civilization."[139] On a reverse tack, Stephen King-Hall expressed concern that a "Regular" IPF—international in its makeup—would be all too vulnerable to a falling-out among the major powers.[140] In the spring of 1943, Winston Churchill, buoyed up by the rising tide of enthusiasm for a strong collective security system, called for "a really effective league, with all strongest forces concerned woven into its texture, with a high court to adjust disputes and armed forces, national or international, or both, held ready to enforce their decisions, and to prevent renewed aggression and preparation of future wars."[141]

While there was considerable substance to some aspects of what was presented as "the special relationship" between the United States and Britain in their wartime virtual alliance, there were also many underlying tensions. Admiral King tried to keep British naval forces out of the Pacific in the last phase of the war, and when Allied naval forces returned to Shanghai at the war's end, the American commander ignored his British counterpart's protests and moored at Number 1 Buoy off the Bund, a space traditionally reserved for the Royal Navy. Some of that was due to longstanding rivalries and hostilities, commercial, ideological, and ethnic. On the other hand, many observers saw the great empires—British, French, Dutch, and Portuguese—as major elements in the world system for the foreseeable future. Some philosophers offered nostalgic praise for the major empires having provided order and security, human costs aside. Salvador Madariaga, for example, on the eve of World War II, called on the British Empire to lead the way to the new postwar world order. George Santayana deemed the British "the sweet, boyish masters of the world," and Reinhold Niebuhr judged Britain's national interest as "more completely identified with the interests of the nations than is the case with the United States." The full extent of the damage done to imperialism by World War II was not apparent.[142] As badly weakened as Britain was, it would keep more forces on active service during the first decade after World War II than the United States. Nevertheless, the mighty surge of American military and naval strength during World War II, well short of the nation's full capacity for mobilization, eased some grassroots anxiety and apprehensions in Congress about the Anglo-American "special relationship," internationalism, and playing in a game where the United States did not hold most of the chips. Many remembered the Allies' vengefulness and rapacity at Paris in 1919, and others distrusted British and French imperialist and/or Soviet designs. But in late 1944, few observers anywhere realized that British voters would soon demonstrate reluctance to bear the burdens of empire by voting Churchill out of office in late July 1945.[143]

There were, however, many signs of a shifting balance of power. Americans' mounting concern over the shape of the postwar world order on the eve of the Dumbarton Oaks meetings was reflected in the sale of over five million copies in five months of Sumner Welles' *Time for Decision*. However, his exit from the State Department in mid-1943 made Welles' monograph less influential in the corridors of power than both an unofficial Brookings "white paper" which offered a refurbished blueprint of Wilson's League with "teeth" and *U.S. War Aims*, a book authored by Walter Lippmann, eminent policy analyst and journalist. In the traditional balance-of-power politics terms favored by American elites, Lippmann dismissed Willkie's *One World*, and Wallace's "Century of the Common Man" as naive. The simplicity and measured tones of *U.S. War Aims* were certainly more attractive to newly appointed Undersecretary of State Edward Stettinius, in its calling for regional spheres of influence dominated by the Big Four. Nodding to "the spectacular development of the striking force of air power," Lippmann depicted North and South America as an "island in an ocean of sea and air"—the geopolitical jargon that became popular during World War II. Condemning the great powers for failing "to take the precaution of becom[ing] members of an indisputably peaceful combination" after World War I, he proposed that while the world was being organized under effective international law, the United States should seek a super-Locarno by forming "dependable alliances in the Old World." Lippmann saw that as leading to a "New Order" based on a permanent core alliance and, most prophetically, as an Atlantic alliance with western Europe in a much diminished status in its overall framework.[144] The Brookings study moved closer to the "bottom line" by judging an IPF as too marginal and expensive on the basis of reviewing the pros and cons of a world security system and tracing the origins of concepts of international order. Most critically, that would require constant patrolling, and bases would have to be secured against insurgent attacks and orchestrated sabotage, both very expensive propositions.[145]

That rising wave of inventiveness and enthusiasm broke against the seawall of diplomatic reality at Dumbarton Oaks in late summer 1944. Detailed proposals for UN forces presented there initially as part of the basic blueprints for the UN organization were prepared by the Advisory Committee on Postwar Foreign Policy, then honed by a group of eight senior State Department officials. Those efforts proved less fruitful than observers or participants expected or appreciated at the time, even though by the time the meeting began, much of the postwar world's structure had already been determined by the flow of military operations. The Russians were moving into eastern and central Europe, where they would remain for almost half a century, Fascist Italy was crushed, and Germany and Japan were crumbling, even though the latter had mortally wounded Western European imperialism in Asia and the western Pacific. China, although gravely wounded, was on a winding and bloody path to resurgence; France was a second-class power, and Britain was farther down that slope than was generally realized at the time. All that became more visible during the Dumbarton Oaks meeting the very moment when the substance of American power deployed in the war matched both the Great Republic's proverbial braggadocio—if not its full productive capacity—and manpower resources. Despite a carefully modulated mobilization and considerable shaving of prewar force plans, the effort was impressive enough. The United States Army Air Forces and Navy were each the largest in the world, and the USAAF's Air Transport Command held a near monopoly on Allied long-range land-based transport aircraft. The Manhattan Project, originally a British and Canadian effort, had been brought under American control, along with the B–29 SILVERPLATE atomic bomb delivery unit, and in September 1944, the number of American ground troops deployed in Europe exceeded the British Empire's for the first time.

Looking back at those events half a century later, it is easy enough to see all the plans for an IPF and IAF, like the League to Enforce Peace, as dreams born of defeat and resolution or as euphoria arising from the prospect of a total victory being won by a collective economic and military effort far greater than anyone imagined. The wartime bloody-mindedness of many of those proposals is also more apparent at a distance. Architects of peace schemes did not anguish over the morality and effectiveness of using air power, or "collateral damage," or the public will. As one IAF enthusiast insisted, "foreign policy is not compounded of public opinion polls taken during the heat of war," an assertion that would be tested frequently after 1945.[146] It is not all that clear, however, what it was that sealed the doom of the IAF and IPF concepts. Clearly, what mattered at Dumbarton Oaks was what the Soviets and Americans were willing to mutually accept. When the Russians arrived, they were genially disposed to a well-structured international order and, like the British and the U.S. State Department, expressed strong interest in forming an IAF. However, they soon shifted their support in favor of FDR's idea of the "Four Big Policeman," even though it is not clear where the President really stood on the IPF/IAF issue at that point. As the Dumbarton Oaks conference began, it was not fully clear that the substantial isolationist bloc was about to go on the rocks in the fall elections. Indeed, many reasonable people feared that agreements made in those meetings might suffer the same fate as those conceived at Paris in 1919. FDR, caught in Wilson's shadow during the 1944 election campaign, may have sensed the danger of locking in on the details of postwar security arrangements too firmly and too soon. He did not like being pinned into corners. Beyond his caution, deviousness, and ambivalence, FDR was, like Wilson, in poor health, and was awash in especially turbulent currents of politics, diplomacy, and public opinion in a war whose outcome was becoming clearer but not absolutely certain. Whatever special sensitivity to international security problems Roosevelt gained from serving in Wilson's government, he learned a lot about Congress during his first two terms and, like most of his generation, had watched the Covenant of the League founder in the U.S. Senate.

Although Roosevelt had gained iconic status internationally and, like Churchill, far more so abroad than at home, so had Wilson. And it was obvious that FDR's line of credit overseas might not be negotiable at home if a bitter battle erupted in the Senate. Although several American presidents had stretched the limits on using military power without a declaration of war, Roosevelt pulled furthest on that tether as commander in chief and head of state. At the same time, his political craftiness and charisma, long tenure in office, and rapidly evolving communications technology produced a unique and powerful combination that generated formidable extremes of adoration and hostility. Amid the preparations for Dumbarton Oaks, American critics of the United Nations sounded much like Wilson's foes a generation before, when they warned that even if Congress approved American membership in an international body, the tempo of crisis and war in the "mechanized age" might move too fast for proper debate and legislative process. The devil in that detail was the fact that making effective provisions for broad contingencies would weaken or short-circuit Congress' constitutionally mandated roles in shaping foreign policy and waging war.[147] That produced the same kind of catch-22 that Wilson confronted. Any agreement struck at Dumbarton Oaks or at a peace conference would have to pass muster before the Senate, but to do that, it would have to be vague and ineffectual.

In the late summer of 1944, although public opinion polls offered strong hints of the impending turn in the political tide in the forthcoming election, isolationism seemed to be in good health. In April, as Stettinius began nudging the State Department's course back toward regionalism, Hull had fallen in line, backing away from his longstanding support for a "superstate with its own police forces and other paraphernalia of coercive power." In a similar spirit, both Democratic and Republican presidential nomination conventions dealt with the international peacekeeping question in what Wendell Willkie called "rubber words." With an Allied victory increasingly likely and coming perhaps sooner than expected, American concerns shifted to postwar economic matters of the kind addressed at the Bretton Woods Conference a month before Dumbarton Oaks. Optimists took heart from that earlier meeting, which did indeed create a more long-lasting framework than the latter, and also from the International Aviation Conference in Chicago, which resolved many chronic problems, like each nation's control over its air space, rights of transit, repair, fueling, and the control of domestic and through traffic.[148] Those complex dynamics had loomed large since the Convention of Paris in 1919, and led various nations to work out unilateral contracts and treaties. By putting those elements of aviation into the framework of a rational international system, in the spirit of the global postal and telecommunications associations, the Chicago conference did much to ease confusion and tensions, drawing the teeth of persistent anxiety over links between civil and military aviation.

Dumbarton Oaks was marked by many proofs of the adage that "In peace the hunger for sovereignty returns"[149] and Hamlet's vision of "the native hue of resolution" being "sicklied o'er with the pale cast of thought." On the other hand, it did not lead to sharp bickering or produce a gloomy sense of failure, although leaks from China's frustrated delegates to the press made it clear enough that the major Allies' "practical" concerns were prevailing over others' visions and hopes. But as at Paris after World War I, widespread perceptions that the war was bringing foreign policy more in line with the public will and farther from traditional diplomacy and great power dominance by creating a fresh, and hopeful international system were proven wrong. Despite many attendees' strong support for IPF and IAF models and widespread popular and official support for those concepts,[150] they had been waylaid within the American defense establishment before the Dumbarton Oaks meetings, most especially by the Special Informal Military Group (SIMG), a team of senior American officers engaged in postwar planning.[151] In a classic Byzantine manner, its functionaries did their work ahead of time and out of view. Earlier, three members had drafted an IPF plan while serving on an advisory committee to the Joint Chiefs of Staff, but they submitted no such proposals in the plan they completed in early November 1943. Counterattacking obliquely, they addressed practical aspects, including base acquisition, of phasing an IPF into service and then concluded there was no way to anticipate what contingencies might lead the United States to deploy forces as part of an IPF.[152] Such subtlety did not effectively smother the proverbial baby. Rather, it aroused FDR's interest in a postwar worldwide network of "air facilities . . . without regard" to sovereignty.[153] At the same time, some State Department officials were attracted to the IAF concept by General Marshall's suggestion that the United States might contribute substantial military forces to an IPF after the war. The diplomats saw the airmen as the most modernist and least burdened by rivalry and traditions among the armed services, while recognizing, as Roosevelt did in his 1940 campaign rhetoric, that using air power to apply military force lay the furthest from the politically unpopular contingency of "sending our boys overseas," that is, in the form of ground troops. The Joint Chiefs of Staff, however, rejected Marshall's proposal in late 1943, and when Dumbarton Oaks convened, Admiral Willson of the SIMG deemed the IPF concept "altogether visionary,"[154] as he and his fellow committee members persuaded Stettinius to take a dim view of such schemes.

The SIMG members, those in the State Department who looked askance at creating a successor to the League,[155] and Stettinius seemed to have their work cut out for them as the Conference began. (Churchill, Roosevelt, and Stalin did not attend, nor did Britain's or the USSR's foreign ministers.) Not only did the British and the Soviets initially support creating a mainly American air contingent serving under Security Council control, but the USSR's design for an "International Air Force Corps" also became the central pivot of discussion at the outset. The Russians felt that "the primary purpose" of the UN was "to establish and maintain international peace and security and that . . . the League of Nations' . . . concentration upon social and economic matters diverted its attention from the enforcement of international peace and security."[156] When they countermarched and fell in step with Stettinius and the SIMG, that left both Churchill, who had overcome his initial doubts, and military professionals and diplomats in the United States and Britain who favored the IPF/IAF approach without effective leverage.[157] FDR, always sensitive to subtle shifts in political tides, now expressed concern that U.S. Army Air Force aircraft might bear United Nations markings—even though he had raised no objection either when RAF roundels were painted on USAAF and Navy antisubmarine warfare aircraft based in the Azores under the Anglo-Portuguese mutual defense treaty or when American volunteer pilots flew Chinese Nationalist planes and served under Dutch, Australian, and British commanders in combat. The message was clear.

Beyond any questions of his personal style or health, Roosevelt's hesitancy made sense, considering the turbulence in the background, which soon included sharp differences between the USSR and the Western Allies over the Polish Home Army's plight in Warsaw. At Dumbarton Oaks, while Britain and Russia opposed China's hopes for a strong centralized model, the United States resisted Soviet and Chinese proposals for creating a standing UN air force comprised of international recruits and opted for on-call contingents—or "quotas." As the Americans realized such a system would put them in the same bind of precommitment that had ensnared Wilson in 1919, the U.S.–Soviet clash over the size of UN forces led Britain to suggest deferring discussion on the matter. Despite its Chiefs of Staffs' willingness to go along with an IPF/IAF, the United Kingdom's senior negotiator, Sir Alexander Cadogan, who privately saw some political value in creating an IAF,[158] joined Stettinius in demanding that the UN Charter provide that requests for military action be approved unanimously by the Security Council members—the five "Great Powers." In time, steering past such organizational details proved the higher realism as the conferees shifted their attention from military structure to the Security Council veto, but it was not fully clear then that the moment for creating an IPF or IAF had passed—if, indeed, that was ever really possible. Like Roosevelt, the Soviets wanted the big powers to dominate the UN, and proposed veto rights to allow any one of them to protect its special interests. The British agreed in broad terms, but preferred a majority vote arrangement that would provide greater freedom to act against aggression. The failure to resolve that issue led Canada's Lester Pearson to label the United Nations another "toothless League,"[159] a more accurate estimate of the situation than most observers realized in late 1944. At the time, however, the stranding of strong advocates of a postwar worldwide democratic order like Republicans Willkie and Stassen, and Democrats Welles and Wallace on the political margins at that point seemed especially ironic since the isolationists had been routed in the fall elections, due in good part to servicemen's absentee ballots. A secret U.S. State Department public opinion poll that showed 80 percent favored using force to maintain peace.[160]

Although some politicians, like J. William Fulbright and Arthur Vandenberg, would prosper from that reversal of fortunes, strong tides soon began to run against the internationalist cause. One stemmed from the declining fortunes of strategic bombing. General James Doolittle, who replaced Ira Eaker as head of the 8th Air Force during the pause in daylight raids in late 1943, ordered fresh attacks against Germany in early 1944, he used mass bomber formations as staked goats to draw Luftwaffe fighters into battle with growing swarms of American long-range escort fighters. In the "Battle of Germany," even though bomber losses fell, fighter losses soared, and modalities and tactics changed, that final struggle in the air, which raged for just over a year, remained a war of attrition and failed to yield a quick, clean decision. Although the USAAF continued to carry out occasional pinpoint attacks, more and more, like the Royal Air Force, it resorted to area bombing, although in daylight. And despite the thinning of its air support, the German Army fought on. In late 1943, half a world away, highly publicized daylight high-altitude attacks on Japan and targets in southeast Asia by B–29s based in southwestern China had also failed to yield clear-cut results. When those raids provoked the Japanese into launching a ground offensive in early 1944, the U.S. 20th Air Force redeployed to the Marianas. As the B–29s moved to Saipan and Tinian in midsummer 1944, a sizable portion of the Allied air forces in Europe were supporting their armies as they grappled with the Wehrmacht in northwest and southern France, Italy, and eastern Europe. In September, the Western Allies' breakout from Normandy had ground to a halt in eastern France and Belgium, while the Anglo–American–Polish airborne thrust into Holland was parried by the Nazis. Although the D-Day air campaign and bombing of oil and transport inflicted deep wounds on the Third Reich, its fighter production soared, and it began to launch *Vergeltung* (vengeance) weapons—the V–1 cruise missiles and V–2 stratospheric rockets—and to deploy jet fighters. Although frightening, such technical wizardry had no decisive effect on overall operations. Mass in the air as well as on the ground, was prevailing over maneuver, bearing out the observations of critics and sentient airmen like Air Marshal Sir John Slessor that air power had become a significant element in the waging of war, but not to the degree that its "prophets" and enthusiasts insisted.

Whatever effect the course of the air campaigns in Europe and the Pacific or other military or naval operations may have had on diplomacy at Dumbarton Oaks, the conference and the UN Charter drafting session in April 1945 in San Francisco produced a shriveled version of the strong, centralized international peacekeeping structure featured in Allied propaganda, official statements, and popular culture throughout the war. Even though the UN Charter's language regarding the use of force against aggression was stronger than the League Covenant's, it became increasingly clear that it was to be far from a world government or police department, but something more along the lines of a loose association, and more like the League than many expected. As the bewildering complications of creating an IPF or IAF came into focus,[161] many enthusiasts still hoped the UNO—as the UN was labeled in the late 1940s—would have standing forces, while others feared—and some hoped—that it, like the League, would become a forum and symbol of lofty ideals, with no standing armed forces, dependent on the great powers for its "teeth," in keeping with FDR's visions of "Four Policemen" and a "posse." On the other hand, there was still enough substance in the UN's design to alarm some members of Congress, who feared that it placed too much discretion in the president's hands,[162] even though the British and Soviets had gone along with Stettinius' preference for UN armed forces comprised of earmarked national contingents, rather than standing multinational forces.

Since Dumbarton Oaks left many details to be worked out among permanent UN Security Council members, the forming of the Military Staff Committee (MSC), to be comprised of the Chiefs of Staff of the Security Council's permanent members, looked like a major step toward giving the UNO a sharp sword. A British proposal, the MSC was meant to be an international general staff of the kind proposed by France and Spain at Paris in the League of Nations era.[163] It proved equally as viable, since, as the United Nations was being formally organized at San Francisco in the spring of 1945, the "Big Three," never as solid a bloc as Allied propaganda suggested, were less and less aligned in policy and purpose. What Churchill deemed "the Grand Alliance" was born of desperate expediency, and as noted earlier, many sharp differences among its members along the way reflected that. It did not collapse all at once, but came apart in pieces, in skirmishes such as wrangles over power sharing in Italy in 1943 and bickering over spheres of influence at Teheran in late 1943 and at Yalta a year later. But the degree of that crazing was not so apparent as it would be later, even though a major crack appeared less than a month after Dumbarton Oaks, when the Soviets urged the non–Communist Polish Home Army in Warsaw to rise against the Germans, stood by as the Wehrmacht crushed it, and refused to cooperate with British and American attempts to provide aid to the insurgents. At that point, Churchill, like many Western observers, saw the Russians' feistiness as the product of their longstanding pariah status and hunger for legitimacy, and was willing to cut them some slack in the interests of avoiding difficulties including the prospect of "World War III."[164] In a similar vein, FDR's public and private support of the "Four Policemen" model became firmer. On the eve of the presidential election of 1944, his views on the postwar world in a radio address sounded much like those he had expressed early in the war, when he said that "the Council of the United Nations must have the power to act quickly and decisively to keep the peace, if necessary."[165] That was hardly surprising, considering Roosevelt's inclination to follow public opinion polls, since Americans strongly favored a powerful United Nations. But differences aside, the major powers seemed to have reached enough of a consensus to allow FDR, in the Christmas radio message he broadcast while the Battle of the Bulge was raging, to put a sharper edge on his rhetoric. He declared that the "Big Four" would "stick together in determination to keep their peace . . . if force is necessary to keep international peace, international force will be applied—or as long as it is necessary."[166] At the same time, Stalin spoke favorably to Churchill regarding the outcome of the Dumbarton Oaks meeting, and the latter discussed IPF structure with Belgian statesman Henri Spaak.[167] Nor did the outbreak of civil war in Greece remove either the working out of the details or the IPF from the agenda when the UN General Assembly convened in San Francisco in April 1945.[168]

Although that two-month long charter drafting session came in the wake of the last and most controversial "Big Three" meeting, the Yalta Conference, the tide of persistent optimism led Edward L. Bernays, a prominent publicist, to wax hopeful on provisions for deploying armed force against aggression developed at Dumbarton Oaks and written into the UN Charter.[169] The noted military journalist, Major George Fielding Eliot, displayed even more chipper optimism in early August in *Look* magazine just before the war ended. Setting forth practical details and estimated costs of America's involvement in policing the postwar world, Eliot envisioned the Royal Navy patrolling "the waters around southern Asia" and a "U.S. naval zone" that "would cover a good two-thirds of the globe." A network of "certain strategic air bases" would support and deploy a 100,000 strong United Nations "Army of Observation" comprised of airborne, armored, and bomber forces, a third of them American.[170] In a less grandiose and more scholarly tone, Grayson Kirk saw Britain, the United States, and the USSR as obligated to use "their strength jointly in behalf of the [United Nations] organization to maintain peace throughout the world," a task requiring "substantial armies, navies and air forces ready for such threats to world peace as may arise." He also pointed out strategists' concern that the British Empire's vast basing system, which the United States and other Allies had benefitted from during World War II, would shrivel away in the postwar era, and those trying to use or retain various modes might well encounter residual antipathy toward imperialism.[171]

Kirk's uncertainty about the future fabric of globalism echoed the sentiments of Stettinius and others of a geopolitical bent, including Sumner Welles, who, as his fortunes declined, had moved far from supporting a global security organization *a la* Wallace and Willkie. Now he envisioned an array of "regional systems of nations charged with the primary duty of maintaining peace . . . under an over-all international body representative of all nations . . . and responsible for the general maintenance of peace."[172] Ironically, just before Stettinius succeeded Hull as Secretary of State, he had been jarred loose from dawdling over the UN Charter draft by internationalists' claims that they could influence four million votes.[173]

That counted for little, however, when Hull gave Stettinius a free hand in reorganizing the State Department, and the latter enhanced the power of its planning subagencies, emphasizing "real world" activities and regional security arrangements.

Neither internationalists' zeal nor the public enthusiasm for a strong international system reflected in polls, and the 1944 American election weighed for much. As the UN Charter's vague provisions took it through the U.S. Senate to ratification, 89 to 2, informal diplomatic assurances and assumptions fed widespread expectations that UN armed contingents would be formed by the Security Council powers in some kind of mix of naval elements from Britain and the United States, ground units from China and Russia, and American air forces.[174] Those intentions and expectations were being swept away by the rapid end of World War II in Europe in May 1945 and in the Far East in August. East-West differences over Poland, Greece, and occupation were unresolved when Roosevelt died in April 1945, a month before the Third Reich collapsed. At that point, War Department planners assumed that "an international organization" created "for the maintenance of world-wide peace and security for the regulation of armaments" would soon be "in full and effective operation" with "each power having primary control in its own strategic areas."[175] Three months later, Stalin stood as the sole survivor of the "Big Three" when Churchill lost the first wartime general election in the middle of the Potsdam Conference.

While the victories both in Europe and the Far East came faster than Allied planners expected, the latter, bathed in the glare of atomic bombs, was far more stunning. The dramatic events of early August 1945 led many observers, including such tough-minded generals as MacArthur and Arnold, to muse on the apparent obsolescence of warfare, but as the sense of shock subsided, it soon became clear that armed conflicts would continue in the nuclear age, although warped into different forms by strange new polarities. The use of nuclear weapons also nearly eclipsed other major factors in Japan's collapse—the Soviet invasion of Manchuria and the USAAF's massive fire bomb raids, similar to Hiroshima and Nagasaki in the scale of destruction. Most important in respect to the IAF concept, after all the setbacks, the most radical air power enthusiasts' vision of air power as the war-winning weapon seemed to have been realized. That was in good part due to the fact that "Hap" Arnold, from the early days of World War II, had struggled to give the U.S. Army Air Forces control over "delivering" atomic bombs, if that technological longshot proved feasible. His efforts also gave the United States a monopoly, however brief that proved to be, and an edge in the long run. America's ascendancy to superpower status was brought home to the world less than a month after Hiroshima, when the surrender of Japan was signed aboard the U.S.S. *Missouri* in Tokyo Bay and over four thousand American warplanes thundered overhead as the ceremony ended. It was, however, not clear if or when nuclear weapons might be used again, and under whose authority.

In the autumn of 1945, amid the first gusts of the Cold War and the massive and chaotic demobilization of America's armed forces, Arnold, now a five-star General of the Army, and near death, mused in *Air Force* magazine on the role of American air power in the postwar era. Considering how to make "the United Nations Charter work," he defined air power as "the means by which we will have to safeguard our country if the United Nations Charter does not work."[176] His reflections had a much keener edge to them than they would have had before Hiroshima and Nagasaki. While Arnold made no mention of USAAF planners' wartime preparations for making American air forces the main element in United Nations peacekeeping forces, that did not matter, and not only because they had been kept away from the business done at Dumbarton Oaks. The structural stability of the postwar order hinged on the concurrence of the Great Powers, which was openly crumbling, taking down with it the public enthusiasm for an international police force and all the heartfelt hopes for a lasting peace that rose to a zenith during World War II.

NOTES

1. E.g., in late October 1939, a British Air Staff Plans Section memo argued that "When we begin to go for Germany properly from the air the rest of the World will heave a sigh of relief." PRO AIR 2/3222/00325, Plans 5, D. L. Blackford, Memo to Deputy Director Plan, October 29, 1939.

2. Allen F. Chew, *The White Death: The Epic of the Soviet-Finnish War* (East Lansing: Michigan State University Press, 1971), p. 127.

3. Raymond Buell, *Isolated America* (New York: Knopf, 1940), p. 456.

4. E.g., Henri Bonnet, *Outlines of the Future: World Organization Emerging from the War* (Chicago: World Citizens Association, 1943), p. 28. Also see F. H. Hinsley, *Power and the Pursuit of Peace: Theory and Practice in the History of Relations Between States* (Cambridge: Cambridge University Press, 1963), p. 315.

5. E.g., Lionel Curtis, *The Way to Peace* (London: Oxford University Press, 1944), pp. 7–9, and Raymond Gram Swing, "The Grass Roots and the Elephants," *The Listener*, November 7, 1943.

6. Its members, who met one Sunday a month, included James Shotwell, veteran crusader for an international system of order, and a senior member of the "Inquiry," the group of academic advisors who advised Woodrow Wilson at the Paris Peace Conference; for details, see Harold Josephson, *James T. Shotwell and the Rise of Internationalism in America* (Rutherford, N.J.: Fairleigh Dickinson Press, 1975), pp. 69–71.

7. *Functions of an International Air Police* [Series F, No. 2] (London: New Commonwealth, 1940), p. 13.

8. Ibid., p. 14.

9. Ibid., p. 89.

10. M. J. Bernard Davy, *Air Power and Civilization* (London: George Allen and Unwin, 1941), p. 146.

11. R. O. MacFarlane, *Beginning at the End* (Toronto: Canadian Institute of International Affairs, 1942), pp. 9–13.

12. W. B. Currey, *The Case for Federal Union* (New York: Penguin, 1939), pp. 33–34. Also see Stuart Chase, "A Policeman for the Planet," Chapter 10, *The New Western Front* (New York: Harcourt Brace, 1939), p. 126.

13. Louise W. Holborn, ed., *War and Peace Aims of the United Nations, September 1, 1939–December 31, 1942* (Boston: World Peace Foundation, 1943), pp. 173–175, 195.

14. E.g., William P. Maddox, *European Plans for World Order* (Philadelphia: American Academy of the Political and Social Sciences, 1940) [James-Patten-Rowe Pamphlet series No. 8]; and Ralph W. Page, "Designs for a World Order," *Annals of the American Academy of the Political and Social Sciences* 210 (July 1940): 50–56.

15. A scheme based on the model of the British Metropolitan Police is Charles Reith, *Police Principles and the Problem of War* (London: Oxford University Press, 1940), esp. pp. 124–144.

16. Freda Kirchwey, "Needed: An American League of Nations," in Egbert Ray Nichols, *Western Hemisphere Defense* (New York: H. H. Wilson, 1941), p. 347.

17. Max Werner, "Postwar Organization of Military Force," *Antioch Review* 1: 2 (Summer 1941): 249–256.

18. William C. Brewer, *Permanent Peace* (Philadelphia: Dorrance and Co., 1940), pp. 29–30.

19. Gerhard Niemeyer, *Law Without Force: The Function of Politics in International Law* (Princeton: Princeton University Press, 1941), p. 389; Niemeyer saw "The Perpetual Demand for International Organization" as "utterly delusional and without foundation," ibid., p. 388. A more recent rigorous analysis of the question is Hidei Suganami, *The Domestic Analogy and World Order Proposals* (Cambridge: Cambridge University Press, 1989).

20. E.g., Robert Corkey, *War, Pacifism and Peace: A Study in War Aims for Peace-Lovers in Britain and America* (London: Society for Promoting Christian Knowledge, n.d., c. 1945), esp. pp. 167–168.

21. Major conclaves and statements of intent included: the "Ten Principles" of Catholic, Anglican, and other Protestant denominations [London, December 1940]; the "Malvern Manifesto" [United Kingdom, January 1941]; the "Delaware [Ohio] Findings" [March 1942]; the "Six Pillars of Peace" [March 1943]; and the "Princeton Messages" [July 1943]; for texts thereof and a very optimistic view on postwar policing arrangements, see Paul Hutchinson, *From Victory to Peace* (Chicago: Willett, Clark and Co., 1943).

22. E.g., the call for "some plan for world federation" in Anna W. M. Wolf, *Our Children Face War* (Boston: Houghton Mifflin, 1942), p. 208.

23. Clarence K. Streit, *Union Now: The Proposal for Inter-Democracy Federal Union* (New York: Harper and Brothers, 1940), pp. 105, 215; for a contemporary critique of Streit that called for a world police force see Buell, *Isolated America*, p. 419.

24. Noel Coward, "Lie in the Dark and Listen," *Atlantic Monthly*, October 1943, p. 98.

25. E.g., RAF Bomber Command included contingents and individual volunteers from Belgium, Czechoslovakia, France, Norway, Poland, the United States, and major Dominions of the British Commonwealth, as well as other nations and colonies.

26. Bonnet, *Outlines of the Future*, p. 36.

27. Sir Charles Webster and Noble Frankland, *The Strategic Air Offensive Against Germany, 1939–1945* (London: H.M. Stationery Office, 1961), 4: 49.

28. The debate is well covered in the official histories, and C. P. Snow's *Science and Government* (Cambridge: Harvard University Press, 1960); for various and divergent views, see *Bomber Command* (New York: Doubleday Doran, 1941); Arthur Harris [Air Chief Marshal Sir, head of RAF Bomber Command] *Bomber Offensive* (New York: Macmillan, 1947), esp. pp. 77–79; P.M.S. Blackett, *Studies of War: Nuclear and Conventional* (New York: Hill and Wang, 1962), pp. 224–225; and PRO AIR 14/3507, Memo, Chief, Bomber Command to Prime Minister, May 30, 1942.

29. In 1944, FDR expressed regrets in private that he had not sent aid to Ethiopia and Spain in the 1930s, cf. Claude H. Eichelberger, *Organizing for Peace: A Personal History of the Founding of the United Nations* (New York: Harper and Row, 1977), p. 239.

30. Julian Huxley, "Armaments and Security," *New Republic* 104:22 (June 2, 1941): 750–753.

31. Holborn, *War and Peace Aims*, pp. 48–51.

32. Benjamin Franklin Trueblood, *The Federation of the World* (Boston: Houghton Mifflin, 1899), p. 131.

33. E.g., Frederick J. Libby, *War on War: Campaign Textbook* (Washington, D.C.: National Council for Reduction of Armaments, n.d. [c. 1923]).

34. Eldon Griffin, *Clinching the Victory* (Seattle: Wilberlilla Publishers, 1943), pp. 39–42.

35. Carl Becker, "How Will the New World Be Better?" *Yale Review* 32:3 (March 1943): 431–432; and Dorothy Thompson, "The Future World Order," *Vital Speeches* 8 (June 15, 1942): 533–536.

36. Walter Lippmann, *U.S. Foreign Policy: Shield of the Republic* (Boston: Little Brown and Company, 1943), pp. 70–76, 121, 126, 161, 175.

37. Henry Bamford Parkes, *The World After War* (New York: Thomas Y. Crowell, 1942), pp. 126, 137–139, 216–232.

38. Robert C. Hildebrand, *Dumbarton Oaks: The Origins of the United Nations and the Search for Postwar Security* (Chapel Hill: University of North Carolina Press, 1990), p. 21.

39. John Steinbeck, *Bombs Away* (New York: Viking Press, 1942), p. 21.

40. "Target for Tomorrow," *Life* 12:19 (May 11, 1942): 12, 14, 15.

41. Quoted in Robert A. Divine, *Second Chance: The Triumph of Internationalism in America During World War II* (New York: Atheneum, 1967), pp. 64–65.

42. E.g., Bernard Davy's assertion that "an international Air Police Force is not at the present stage of evolution desirable" nor feasible in the near future, in *Air Power and Civilization*, p. 196.

43. Quoted in Philip Curtis Nash, *An Adventure of World Order* (Boston: Beacon Press, 1944), p. 19; also see Eichelberger, *Organizing for Peace*, p. 241.

44. Quoted in Allan A. Michie, *Keeping the Peace Through Air Power* (New York: Henry Holt, 1944), p. 119.

45. Holborn, *War and Peace Aims*, pp. 234, 395–396.

46. Evan Luard, *A History of the United Nations*, vol. 1, *The Years of Western Domination 1945–1955* (New York: St. Martin's Press, 1982), pp. 19–20.

47. *The Autobiography of James T. Shotwell* (Indianapolis: Bobbs-Merrill, 1961), p. 312.

48. Divine, *Second Chance*, pp. 66–71; also see Holborn, *War and Peace Aims*, p. 652.

49. Henry A. Wallace, "Address at the Congress of American-Soviet Friendship Celebrating the 25th Anniversary of the Founding of the Soviet Union," in Holborn, *War and Peace Aims*, p. 134.

50. Michael Sherry, *Preparing for the Next War: American Plans for Defense, 1941–1945* (New Haven: Yale University Press, 1977), p. 42.

51. Sumner Welles, "U.S. State Department Memorial Day Address at Arlington National Amphitheater, May 30, 1942," in Holborn, *War and Peace Aims*, p. 88.

52. Liberal Party on International Affairs, Submitted by the Executive to the Assembly of the Liberal Party at Caxton Hall, London, September 1942, in Holborn, *War and Peace Aims*, p. 685.

53. Nicholas Doman, *The Coming Age of World Control* (New York: Harper and Brothers, 1942), pp. 155, 211.

54. Burnet Hershey, *The Air Future: A Primer of Aeropolitics* (New York: Duell, Sloan, and Pearce, 1943), pp. 161, 173–174.

55. George T. Renner, "How the United Nations Must Police the World," *American Magazine* 136:9 (September 1943): 31.

56. Michie, *Keeping the Peace*, p. 163. For another version of the airborne army IPF model, see Volta Torrey, "Will Airborne Police Enforce World Peace?" *Popular Science* 145:3 (September 1944): 72.

57. E.g., Jackson H. Ralston, *A Quest for International Order* (Washington, D.C.: John Byrne, 1941); and *World Organization—A Balance Sheet of the First Great Experiment* (Washington, D.C.: American Council on Public Affairs, 1942), p. 399.

58. Norman Thomas, *World Federation: What Are the Difficulties*? (New York: Postwar World Council, 1942), pp. 6–17.

59. J. C. Smuts, *Plans for a Better World* (London: Hodder and Stoughton, 1942), pp. 119–120, 280.

60. Allied forces in the Mediterranean campaigns included forces from Britain, all of the Commonwealth's Dominions, the United States, metropolitan France and its Empire, Brazil, Greece, Poland, and Palestine; the Waffen SS included elements from France, Belgium, Holland, Scandinavia, and Balkan ethnic groups; Bulgarian, Hungarian, Rumanian, Italian, Spanish and Finnish forces served under Wehrmacht command on the eastern front; and Chinese, British, American, West African, Australian, New Zealand and Indian troops served in Burma.

61. For details, see Max Werner, "Mastering the Military Machines," in J. Donald Kingsley and David Peregorsky, eds., *Strategy for Democracy* (New York: Longmans, Green & Co., 1942), pp. 80–81.

62. Speech to a group of clubs, Chicago, December 16, 1942, quoted in Holborn, *War and Peace Aims*, p. 665; and Herbert Hoover and Hugh Gibson, *The Problem of Lasting Peace* (New York: Doubleday Doran, 1942), esp. pp. 257–262.

63. Scott Nearing, *The Tragedy of Empire* (New York: Island Press, 1945), pp. 161–162.

64. E.g., Eberhard Billeb, "The Airplane as a Geopolitical Force," in Andrew Dorpalen, ed., *The World of General Haushofer: Geopolitics in Action* (New York: Farrar and Rinehart, 1942), p. 134.

65. "Simplicus and Perplexides on the Use of Force to Insure Peace," *School and Society* 56 (August 29, 1942): 165–166; and 57 (March 6, 1943): 268–269.

66. E.g., the dependence of an international force on "intermediate steps of moral compulsion, imposed by self-assured obligation—by a promise."

67. E.g., see Murray Harris, *The Logic of War* (London: George Allen and Unwin, 1944), pp. 125–129.

68. Other American "tough guy" naval and Marine commanders were Marine generals Arthur Vandegrift and "Howlin' Mad" Smith, and admirals Richmond Kelly "Terrible" Turner, William F. "Bull" Halsey, and Marc Mitscher.

69. E.g., the USAAF's Lewis Brereton, Ira Eaker, Curtis LeMay, and Claire Chennault.

70. For a contemporaneous overview, see Buel W. Patch, "Enforcement of World Peace," *Editorial Research Reports*, October 30, 1941, pp. 299–302.

71. Currey, *The Case for Federal Union*, pp. 33–34. Also see Chase, "A Policeman for the Planet," p. 126.

72. Hardrop Freeman and Theodore Paullin, *Coercion of States: In Federal Unions* (Philadelphia: Pacifist Research Bureau, 1943), p. 60.

73. For details, see Hildebrand, *Dumbarton Oaks*, pp. 24–26.

74. Sumner Welles, *The World of the Four Freedoms* (New York: Columbia University Press, 1943), p. 119, and *The Time for Decision* (New York: Harper and Brothers, 1944), pp. 380–382.

75. Martin Gilbert, *Winston S. Churchill* (Boston: Houghton Mifflin, 1986), 7:493.

76. Perhaps the most critical analysis is Charles A. Beard's *President Roosevelt and the Coming of the War: A Study in Appearances and Realities* (New York: Archon Books, 1960), pp. 465–478.

77. Herman S. Wolk, *Planning and Organizing the Post War Air Force, 1943–1947* (Washington, D.C.: Office of Air Force History, 1984), pp. 46–50, 227–236, 302, and *The Struggle for Air Force Independence* (Washington, D.C.: Air Force Historical and Museum Programs, 1982), pp. 52, 254.

78. Ely Culbertson, *Total Peace* (New York: Garden City Publishing, 1943), esp. pp. 23–28, 180, 245, 274–278. For a condensed version of the plans, see Culbertson's *Summary of the World Federation Plan* (New York: World Federation, 1943), and for basic elements, see his "System to Win this War and the Peace to Come," *Reader's Digest* 42 (February 1943): 135–142.

79. Culbertson, *Summary of the World Federation Plan*, p. 71.

80. Michie, *Keeping the Peace*, p. 115.

81. Robert Rhodes James, ed., *Winston S. Churchill: His Complete Speeches 1897–1963* (New York: Bowker, 1974), 7:6778.

82. Ibid., p. 6798.

83. Hildebrand, *Dumbarton Oaks*, p. 23.

84. George H. Gallup, ed., *The Gallup International Public Opinion Polls—Great Britain 1937–1975* (New York: Random House, 1976), p. 96; Jerome S. Bruner, "Public Opinion and the Peace," in J. B. Whitton, ed., *The Second Chance: America and the Peace* (Princeton: Princeton University Press, 1944), pp. 185–187; Divine, *Second Chance*, p. 110.

85. Henry A. Wallace, "What We Will Get Out of the War," *American Magazine* 85:3 (September 1943): 98.

86. "The Organization of the Peace: An Outline of the New Commonwealth Program," *The Fighting Forces* 21:5 (December 1944): 257–261.

87. Henry M. Wriston, *Strategy of Peace* (Boston: World Peace Foundation, 1944), p. 67.

88. And a book title—Michael Straight's *Make This the Last War* (New York: Harcourt Brace, 1943).

89. Louis Fischer, "Peace, the Culbertson System," *Nation* 156:17 (April 24, 1943): 582, and Thompson, "Future World Order," pp. 533–536.

90. Divine, *Second Chance*, p. 109.

91. Joseph H. Ball, *Collective Security: The Why and How* (Boston: World Peace Foundation, 1943), p. 35–44.

92. Senate/House of Representatives Resolution 114–1/78—March 16, 1943; Ball was a Democrat, as were sponsors Carl Hatch, New Mexico, and Lister Hill, Alabama; Harold Burton of Oklahoma was a Republican.

93. E.g., see *Should There Be an International Organization for General Security Against Military Aggression, and Should the United States Participate in Such an Organization?* [Problem IV] (Boston: Universities Committee on Post-War International Problems, 1944).

94. A detailed discussion of dilemmas in international law and diplomacy linked to an IPF was R. M. McIver, *Toward an Abiding Peace* (New York: Macmillan, 1943), esp. pp. 29–39, 122–123.

95. W. C. Allee, "Where Angels Fear to Tread: A Contribution from General Sociology," *Science*, June 11, 1943, p. 524.

96. *The Autobiography of James T. Shotwell*, pp. 310–312, and James T. Shotwell, *The Great Decision* (New York: Macmillan, 1944), p. 124.

97. Luard, *History of the United Nations*, 1:24.

98. Clyde Eagleton, "Why Not an International Police Force?" *Scholastic* 33 (December 17, 1943): 145ff.

99. E.g., the cautious mini-essays by Reinhold Niebuhr, Grayson Kirk, and G. Bromley Oxnam on "Foreign Policy and World Organization" in *Nation* 159 (October 21, 1944): 489–491; also see Ruth Bryan Owen, *Look Forward, Warrior* (New York: Dodd, Mead, and Company, 1952).

100. James T. Patterson, *Mr. Republican: A Biography of Robert A. Taft* (Boston: Houghton Mifflin, 1972), p. 291.

101. E.g., for an overview, see Divine, *Second Chance*, pp. 32–38, 83–89.

102. *Building Peace: Reports of the Committee to Study the Organization of Peace, 1939–1972*, (Metuchen, N.J.: Scarecrow Press, 1973), esp. "The Organization of an International Air Force," 1: 116–124.

103. Hildebrand, *Dumbarton Oaks*, p. 140.

104. Ibid., p. 143.

105. Webster and Frankland, *Strategic Air Offensive*, 1: 297–298.

106. Louis P. Lochner, ed. and trans., *The Goebbels Diaries* (New York: Eagle Books, 1948).

107. William Roger Louis, *Imperialism at Bay: The United States and the Decolonization of the British Empire* (New York: Oxford University Press, 1978), p. 263.

108. JCS 570, cited in Sherry, *Preparing for the Next War*, pp. 43–44.

109. For details, see Perry McCoy Smith, *The Army Air Force Plans for Peace, 1943–1945* (Baltimore: Johns Hopkins University Press, 1970), pp. 43–49.

110. For Vandenberg's animadversion as part of the transition to "bi-partisan foreign policy," see C. David Tompkins, *Senator Arthur H. Vandenberg: The Evolution of a Modern Republican 1884–1945* (East Lansing: Michigan State University Press, 1970).

111. Patterson, *Mr. Republican*, pp. 286–291.

112. E.g., see J. Eugene Harley, *Woodrow Wilson Still Lives—His World Ideals Triumphant* (Los Angeles: Center for International Understanding, n.d. [c. 1944]).

113. Julia E. Johnsen, ed., *Reconstituting the League of Nations* (New York: H. H. Wilson, 1943); and idem, *International Police Force* (New York: H. H. Wilson, 1944).

114. E.g., J. B. Whitton, *The Second Chance: America and the Peace* (Princeton: Princeton University Press, 1944), pp. 102–108.

115. American Historical Association, *Can We Prevent Future Wars?* [War Department Education Manual—EM 12. G.I. Roundtable] (Madison, Wis.: USAFI, 1944).

116. Carl Becker, *How New Will the Better World Be? A Discussion of Postwar Reconstruction* (New York: Alfred A. Knopf, 1944), pp. 175–180.

117. Charles Beard and Mary Beard, *A Basic History of the United States* (New York: Doubleday Doran, 1944), p. 481.

118. Harry Lewis Braham, *A Permanent Peace for Europe* (Boston: Christopher Publishing House, 1994), pp. 171, 186.

119. Dexter Perkins, *America and Two Wars* (Boston: Little Brown, 1944), p. 100. Also see Quincy Wright, "The Coronet Round Table: Will There Be Another War?" *Coronet* 14:5 (September 1943): 177.

120. *Should There Be an International Organization for General Security Against Military Aggression, and Should the United States Participate in Such an Organization?*, pp. 2–7.

121. For an insurance scheme to compensate "tradal [*sic*] and industrial groups" for "sacrifices that they themselves must make" in applying sanctions, see John H. Wigmore, "Bullets or Boycotts: Which Shall be the Measure to Enforce World Peace?" *American Bar Association Journal* 29 (September 1943): 491–493.

122. Mortimer Adler, *How to Think About War and Peace* (New York: Simon and Schuster, 1944), pp. 18–19.

123. Clyde Eagleton, *Forces That Shape Our Future* (New York: New York University Press, 1945), pp. 108–117.

124. P. E. Corbett, *Post-War Worlds* (New York: Farrar and Rinehart, 1942), p. 137.

125. Richard N. Coudenhove-Kalergi, *Crusade for Pan-Europe: Autobiography of a Man and a Movement* (New York: G. P. Putnam's, 1943), pp. 294–303.

126. Bonnet, *Outlines of the Future*, p. 55.

127. Quoted in Donald Bennett, *Freedom from War* (London: Pilot Press, 1945), p. 29.

128. William Beveridge, *The Price of Peace* (New York: W. W. Norton, 1945), esp. pp. 79–82.

129. W.J.R. Fox, *The Super-Powers: The United States, Britain, and the Soviet Union—Their Responsibility for World Peace* (New York: Harcourt and Brace, 1944), pp. 9, 155.

130. Bennett, *Freedom from War.*

131. William A. Bishop, *Winged Peace: The Air Age* (Toronto: Macmillan Co. of Canada, 1944), p. 173.

132. E.g., for a British civil servant's proposal for a somewhat Kiplingesque "Air Board of Control," see Frank Stuart, "Let's Keep the Air Free," *Coronet* 15: 7 (May 1944): 8–12.

133. Alaska, American Samoa, the Canal Zone, Guam, Hawaii, Johnston Island, the Philippines, Puerto Rico, the Virgin Islands, and Wake Island.

134. E.g., Lionel Gelber, *Peace by Power* (London: Oxford University Press, 1942), pp. 7–8.

135. Nicholas Murray Butler, "The American Plan to Prevent War," *Vital Speeches* 8 (June 15, 1942): 520–521.

136. E.g., see Allan A. Michie, "Air Control Without Occupation," in *Keeping the Peace*, pp. 137–158.

137. Alden P. Armagnac, "What's Being Planned for Our Postwar Battle Strength," *Popular Science* 145:2 (August 1944): 244.

138. Lord Davis, *The Seven Pillars of Peace* (London: Longmans, Green and Co., 1945), pp. 57–92.

139. Norman Angell, *Let the People Know* (New York: Viking, 1943), pp. 50–51, 58; also see A. E. Farncombe and C. A. Pana, *World Government: Shall Britain Participate?* (London: British-American Council for World Government Organization, 1944).

140. Stephen King-Hall, *Britain's Third Change* (London: Faber and Faber, 1943), p. 121; detailed plans included Lionel Curtis, *World War: Its Cause and Cure* (London: Oxford University Press, 1945), and Victor Altman, *The International Police and World Security* (London: Alliance Press Ltd., 1945), esp. pp. 70–82.

141. Quoted in Altman, *International Police*, p. 8.

142. Reinhold Niebuhr, *The Children of Light and the Children of Darkness: A Vindication of Democracy and a Critique of Its Traditional Defense* (New York: Scribner's, 1944), p. 184.

143. Fox, *The Super-Powers.*

144. For general background, see Lippmann, *U.S. Foreign Policy.*

145. Arthur C. Millspaugh, *Peace Plans and American Choices* (Washington, D.C.: Brookings Institution, 1942), pp. 89–99.

146. Michie, *Keeping the Peace*, p. 115.

147. For perspectives, see John Grafton Rogers, *World Policing and the Court: An Inquiry into the Powers of the President and Congress—Nine Wars and a Hundred Military Operations, 1789–1945* (Boston: World Peace Foundation, 1945), pp. 86–89; Hildebrand, *Dumbarton Oaks*, p. 150; and Georg Schild, *Bretton Woods and Dumbarton Oaks: American Economic and Political Postwar Planning in the Summer of 1944* (New York: St. Martin's Press, 1996), pp. 144–161.

148. A participant's perspective at the end of the twentieth century is L. Welch Pogue, "The Next Century of Flight," *Aviation Week and Space Technology*, February 14, 2000, p. 90.

149. See "The 'Fraternal Association,' " *Fortune* 30:1 (July 1944): 115–118.

150. For details on General Marshall and U.S. Army Air Force leaders' views, see Herman S. Wolk, *The Struggle for Air Force Independence, 1943–1947* (Washington, D.C.: Air Force History and Museum Program, 1982), pp. 52, 254, and 262. A list of basic IPF models drawn up by an RAF officer on the eve of Dumbarton Oaks included:

1. Britain and its major dominions (obsolete)
2. a Trustee system, based on contingent forces of the Big Four (the United States, the USSR, Britain, and China), and Axis nations disarmed (which he foresaw as generating an arms race)
3. a Quota system, based on the old French-New Zealand plan, with on-call national contingents established in various countries but committed under their own commanders
4. a Quota system under an International/U.N. staff and commander
5. independent, international forces directly recruited and under an International Staff (which he deemed "the ultimate ideal solution")

151. Chair, Vice Admiral Russell Willson; members, the Army's George V. Strong (Lt.–Gen.), Stanley Embick (Lt.–Gen.), and Army Air Forces' Muir S. Fairchild (Maj.–Gen.).

152. *Postwar Foreign Policy Preparation 1939–1945* (Washington, D.C.: U.S. Department of State, 1949), pp. 76–77; Schild, *Bretton Woods and Dumbarton Oaks*, p. 147; Herbert Feis, *Churchill, Roosevelt, Stalin: The War They Waged and the Peace They Sought* (Princeton: Princeton University Press, 1957), p. 108.

153. Louis, *Imperialism at Bay*, pp. 259–261.

154. Ibid., p. 372.

155. E.g., see George Kennan, *Memoirs, 1925–1950* (Boston: Little Brown, 1967), pp. 217–219.

156. Hildebrand, *Dumbarton Oaks*, p. 145.

157. See National Archives Records Group 218, Records of the U.S. Joint Chiefs of Staff, Chairmans' File, Admiral Leahy, 1942–1948, Box 19, Folder 117, *Army and Navy Members of the American Group, Four Power Discussions Memorandum for Information No. 2*, August 17, 1944, pp. 1, 2; for a Cold War perspective, see Philip Noel-Baker, *The Arms Race: A Program for World Disarmament* (New York: Oceana Publications, 1958), p. 550.

158. David Dilks, ed., *The Diaries of Sir Alexander Cadogan, O.M., 1988–1945* (New York: G. P. Putnam's, 1972), p. 660.

159. National Archives Records Group 218, Records of U.S. Joint Chiefs of Staff, Central Decimal File, Box 30, Folder CCS 092, April 14, 1945, Secs. 1–14, "Guidance for the U.S. Representatives on the Military Staff Committee of the United Nations," pp. 151–153.

160. Divine, *Second Chance*, p. 252.

161. Hildebrand, *Dumbarton Oaks*, p. 147.

162. Ibid., p. 149.

163. Ibid., pp. 156–158.

164. Francis L. Loewenheim, Harold D. Langely, and Hanford Jonas, *Roosevelt and Churchill: Their Secret Wartime Correspondence* (New York: Saturday Review Press, 1975), pp. 577–578.

165. Quoted in Richard Koebner and Helmut Dan Schmidt, *Imperialism: The Story of a Political World, 1840–1960* (Cambridge: Cambridge University Press, 1964), p. 333.

166. Quoted in Altman, *International Police*, p. 50.

167. Loewenheim, Langley, and Jonas, *Roosevelt and Churchill*, p. 584.

168. See *Documents of the United Nations Conference on International Organizations* (London: United Nations Information Organizations, 1945), 3:386–387, 12: 391, 552, 583, 610–613.

169. Edward L. Bernays, *Take Your Place at the Peace Table: What You Can Do to Win a Lasting Peace* (New York: Gerent Press, 1945).

170. George Fielding Eliot, "What Will It Cost Us to Police the World?" *Look*, August 7, 1945, p. 82; a British variant, with longer teeth, is Captain B. A. Stuart's "An Antidote to Aggression," *The Fighting Forces* 22:3 (August 1945): 159–161.

171. Grayson Kirk, "Security for the United States: How Can We Achieve It?" in Harold Sprout and Margaret Sprout, eds., *Foundations of National Power* (Princeton: Princeton University Press, 1945), pp. 767–768.

172. Sumner Welles, ed., *An Intelligent American's Guide to the Peace* (New York: Dryden Press, 1945), p. 7.

173. *Autobiography of James T. Shotwell*, p. 313.

174. Chapter VII, Articles 42, 43, and 45, *United Nations Conference on International Organization, Facsimile of the Charter of the United Nations* (Washington, D.C.: U.S. Department of State, 1945), p. 9.

175. Wolk, *The Struggle for Air Force Independence*, p. 262.

176. Henry H. Arnold, "Our Power to Destroy War," *Air Force* 28:10 (October 1945): 36.

V

Blossoms Scattered and Faded: Collective Security since World War II

As the United Nations Charter was being drafted at the San Francisco conference in early 1945, a Mexican delegate foresaw that it would produce "a world order in which the mice could be stamped out but in which the lions would not be restrained."[1] Events over the next half century bore out his prediction; a great many "mice" were stomped in over a hundred major conflicts and a host of minor ones, leaving about twenty million dead in their wake.[2] But at least the lions, bears, and eagles did not go all out at each other, and there was not another World War. In contrast with the League of Nations era, many multinational military forces were organized during the Cold War to "wage peace,"[3] and air power played a significant role in most of the "hot" conflicts of that era, from small insurgencies to full-blown conventional wars.[4] Although troops, ships, and aircraft under United Nations auspices were frequently engaged, widespread hopes in America, Britain and elsewhere during World War II that standing forces would serve under the UN banner were confounded.

The pattern of global conflict during the last half of the twentieth century led a veteran of many UN struggles to exclaim "How right he was!" when looking back at geographer Nicholas Spykman's 1943 warning that "plans for far-reaching changes in . . . international society," as "intellectual byproduct[s] of all great wars," had "never altered fundamental power patterns."[5] Other seers of that era also got it right, including those who expected the Military Staff Committee to be a main point of contention between the big powers, since the MSC most specifically embodied assigning control of military forces to a supernational entity.[6] The situation was complicated by the fact that as the euphoria of victory ebbed rapidly in the late 1940s, so did the brief spasm of hope that nuclear weapons would make war extinct. Not only did lengthy and complex haggling between East and West include conflicts over creating a formal structure to enforce peace, but enough wars had erupted soon after World War II to puncture the bubbles of optimism. In the American election campaign of 1948, even Henry Wallace, now on the far Left as Progressive Party presidential candidate, played the IPF theme softly, even more so than in his final days as vice president.[7] On the sidelines, military theorists like B. H. Liddell Hart and Otto Heilbrunn observed how the powerful political and military gravity field of "nukes" was forcing international conflict into stunted forms like bonsai trees. On the one hand, the likelihood of major clashes between the superpowers and/or their major client states was reduced by soaring potential costs of accidents and subtle provocations, as well as those of a general war, but, on the other, the likelihood of "limited" and surrogate wars seemed to be increasing.

Over time, the tangle of postwar conflicts arising from superpower animus and decolonization crushed plans for a structured postwar international order, making diplomacy and military policy in the Cold War even more bewildering than they had been during the imperial era. But despite increasing American-Soviet estrangement, prospects for creating some kind of IPF under UN auspices did not crumble immediately after World War II. "Nukes" did not immediately displace conventional strategic bombing in American war planning, nor did they dull the enthusiasm of collective security architects. Nor, in spite of the "V" weapons, did rockets. As it took almost two decades for them to displace aircraft as the principle means of wielding strategic nuclear power, the shrinking cohort of IAF enthusiasts continued to see multiengine long-range heavy bombers as the main modality of imposing peace through force. The model of a policeman with a sledgehammer had been hard enough to sell; a cop driving a bulldozer was out of the question. It was also increasingly clear that nuclear weapons had far less utility in waging war than they seemed to have in the immediate aftermath of Hiroshima and Nagasaki. They quickly became pejorative symbols, and as the superpowers began building large nuclear arsenals, President Truman was accused of using the atomic bomb against Japan to threaten the Soviet Union as well as to demonstrate both its power and America's willingness to use them. As East-West tensions heightened, various attempts to put the atomic genie back in the bottle failed. In the spring of 1946, the Soviets rejected the proposal by Bernard Baruch, America's ambassador to the United Nations, to place nuclear research and weaponry under an international agency. At the same time, Truman made a veiled threat to use atomic weapons to prevent a Russian occupation of Iran in the late summer, in the wake of Winston Churchill's "Iron Curtain" speech. Congress imposed an American monopoly over development and control of atomic bombs and nuclear power by creating the Atomic Energy Commission, brushing aside Roosevelt and Churchill's agreement at Hyde Park in 1944, that the Anglo–American collaboration in the Manhattan Project would continue after the war. On the other hand, it also authorized American military involvement in peace-keeping operations by passing the United Nations Participation Act.[8]

Despite rising Cold War tensions, plans for enforcing world order with military force continued to appear in the American media. For example, the prominent American military journalist George Fielding Eliot, who claimed close ties to senior military planners, presented an expanded vision of "police operations." It was based on Roosevelt's "four policeman" model and the assumption that superpower tensions would soon be resolved. Under Eliot's command structure, which resembled the wartime U.S.–British Combined Chiefs of Staff, enforcers would brandish or use conventional and, in some cases, obsolescent weapons to cow or defeat poorly armed dissidents, as the U.S. Navy's special squadron in Latin America and the navies of many Western nations did in China between the world wars. Despite his concern that military leaders might confuse the relative ease of deploying force in relatively marginal operations with its declining utility in larger wars, Eliot, like many policymakers and analysts from World War II to the Korean War, saw Anglo-American naval elements as the "ready forces" mentioned in Article 43 of the United Nations Charter and saw UN ground forces as Russian.[9]

A more intricate and apocalyptic vision was offered by historian Edward Mead Earle, in the manner of late nineteenth-century naval theorist Alfred Thayer Mahan's *The Influence of Sea Power Upon History*. In his essay entitled the "The Influence of Air Power Upon History," Earle defined World War II as a case of "modern war" evolving into "war to the death against cities" and making "air power" one "of the most pressing and inescapable problems of modern statecraft." From his perspective, half a decade after Pearl Harbor, eighteen months after Hiroshima, and four years before the Soviets tested their first atomic bomb, Earle anticipated a nuclear "knockout blow." Defining "all industry" as "war industry," he urged the dispersal of "industry, transportation and government" and warned that Britain, "so long an outpost of our security, may well be indefensible" in the future.[10] After concluding that "unless we destroy war, war will ultimately destroy us," Earle insisted that the assumption that war was inevitable be discarded since "ultimate security lies in the conquest of war," not in trying to limit armaments.[11]

A similar earnestness flavored Cord Meyer Jr.'s proposal a year later.

A militant but hard-headed pacifist, who later left that path and rose to high rank in the CIA,[12] Meyer urged amending the UN Charter to give it sharper "teeth" to enforce international law. His design looked much like French and Spanish interwar models and many of the designs that appeared during World War II. Its central element was a worldwide federal system under the UN with its "own police and military forces," network of bases, and research and development establishment, which would carry out arms control inspection and enforcement. Meyer proposed that the military elements owe "exclusive allegiance to it" and that it be "sufficiently armed" to promptly "execute the world laws" and comprised of mixed nationalities recruited only from the democracies.[13] But despite such persistent hopefulness, it became increasingly clear in the late 1940s that the United Nations was a coordinative bureaucracy and forum for debate much like the League of Nations. Assailing sovereignty as the root of the UN's lack of "teeth," Emery Reves identified the crucial pitfall as continuing adherence to a "Ptolemaic" principle of self-determination in a "Copernican" world of modern international relations, in which interdependence was the fundamental reality.[14]

Although some skeptics argued that such schemes were as naive as they had been in Dante's time, IPF/IAF blueprints that appeared immediately after the war were taken more seriously in the realm of power politics than they would be a generation later.[15] The UN General Assembly, for example, approved a list of disarmament statements that included a mandate for creating "special organs" and recommended that "the Security Council . . . accelerate as much as possible the placing at its disposal of the armed forces mentioned in Article 43 of the Charter."[16] France's new constitution provided for limiting its sovereignty on a reciprocal basis, a faint echo of Churchill's 1940 offer of joint citizenship to France, while Canadian wartime prime minister Mackenzie King called for "some surrender of national sovereignty" in "a world order under a rule of law," as Canada configured its armed forces to serve as a UN contingent.[17]

In the end, however, all those ascending paths ran into cul-de-sacs. Attempts by the UN's first secretary general, Trygve Lie, to create an "armed forces of the Security Council" led nowhere,[18] and for the next four decades, all UN force deployments would be reactive, comprised of national contingents, and paid for mainly by the major powers. Along the way, the term "police" would become blurred, used to describe actual local security and maintenance of law and order in some instances, while referring to the use of full-scale military power in others.

As American public and elite enthusiasm for substantial IPF/IAF models ebbed, former U.S. Secretary of State James Byrnes blamed the failure to form a permanent UN force on "the delaying tactics of the Soviet representative on the Military Staff Committee of the Security Council" and "the Soviet Union's reluctance to accept the international controls and the delegation of national sovereignty that the overwhelming majority of nations believe are essential."[19] But despite those contrary trends, USAAF and then United States Air Force (USAF) planners continued to prepare for the IAF contingency long after World War II, while civilian proponents of a strong air force cited German commanders, survivors of Allied air attacks in World War II, who urged placing "a strong air force . . . at the disposal of the [UN]" to "put down" or "delay any unfriendly action" to allow "other . . . means to be employed."[20] As noted earlier, General Arnold, in his final days as head of the Army Air Forces in late 1945, anticipated the UN's failure. That theme was repeated by his successor, General of Army Carl Spaatz, last head of the USAAF and first Chief of Staff of the independent U.S. Air Force created in 1947. Decoupling American airpower from an internationalist peace-enforcing role, Spaatz urged creating a strategic bomber-based retaliatory force, making no reference to the United Nations or collective security.[21] This plan was consistent with his vision of a future major war whose outcome "would probably be decided by some form of air power before the surface forces were able to make contact with the enemy in major battles."[22] That blurring of structure and purpose reflected the central problem encountered by American military planners, as well as analysts and pacifists, as they drew plans for UN armed forces—the uncertainty of where nuclear weapons fit into military doctrine and practice. In the late 1940s, Bernard Brodie and other defense theorists identified several linked dilemmas, including the vindictive and self-destructive impulses lying at the heart of deterrence, the suppressing effect of the nuclear standoff on international efforts to thwart aggression, the destabilizing effects of technical evolution, the growing likelihood and soaring potential costs of tactical and strategic surprise, and increasingly intricate problems of protecting military forces, bases, and cities from nuclear weapons' effects.[23] Nor was it clear how much the Soviets acted under their own hidden logic in preparing for nuclear war, rather than conforming to American thermonuclear theories, assumptions, and doctrines.[24] The conceptual maze of nuclear theoretics led RAF Group Captain Fulljames, tireless designer of intricate IAF schemes in the 1930s and during World War II, to return to his drawing board. Recognizing that the veto power of the major nations on the Security Council had rendered "largely illusory the establishment of an international Police Force," he presented a modified version of his earlier model of a core of multinational standing forces serving under United Nations command and added strong intelligence and inspectorate elements, whose task would be the rooting out of atomic weapons.[25] Veteran British air power theorist J. M. Spaight proffered a sharper-edged model. Pointing out that it was taking years to rebuild the cities that had been devastated by bombing in World War

II, he suggested that a United Nations air force should be created to serve as an "international fire brigade," which would bomb key "transportation and oil targets" while sparing population centers.[26]

Traces of interwar French IPF proposals were visible in veteran hard-line pacifist Grenville Clark's concept of "qualitative disarmament," an idea based on his work on postwar planning for the U.S. State Department during World War II. In 1950, on the eve of the Korean War, he co-authored a landmark IPF design with Louis Sohn, a Harvard Law School faculty member. In proposing that a "United Nations Peace Force" be linked with a judicial and conciliation system, which would be created after general disarmament, they aligned their design with official American statements of interest, including those of the U.S. ambassador to the UN Ralph Bunche.[27] In its basic structure, the Clark-Sohn plan strongly resembled 1920s and 1930s disarmament conference agendas in denying all nations tanks, military plans, bombs, artillery, and warships and in proposing an end to military industrial production. Aside from a residue of "lightly armed local and national police for the sole purpose of internal order," only small national contingents dedicated to a United Nations Peace Force would be maintained, following a twelve–year stepped reduction under close UN scrutiny. The Security Council was to be replaced by an Executive Council, which would oversee a committee of five commanders. In spite of its rigor and elegance, the Clark–Sohn plan proved no more practicable than all the others, despite a brief flurry of interest in official circles in the United States and Soviet Union in the early 1960s.[28]

Although the Clark-Sohn design impressed many hard-line pacifists,[29] as well as some policymakers, the diplomatic soil proved as barren for the flourishing of such schemes as it had been in the League of Nations era. By 1948, the Military Staff Committee of the United Nations had been able to reach a consensus on only half the issues presented to it and on no major question. That failure reflected the chronic lack of agreement among permanent Security Council members regarding such matters as the relative size of contingents, basing, and the actual size of the overall force. As the MSC's futile deliberations dashed hopes for even a very loose world federation,[30] its meetings continued during the five years following World War II. As superpower rivalry produced an impasse and nearly paralyzed the UN General Assembly process, regional wars of varying duration and intensity erupted in Indonesia, Palestine, Kashmir, and China. Like nearly all the "shooting wars" of the Cold War era, they were products of decolonization or its aftermath in Asia and Africa. Despite all the "limited wars," "unconventional wars," and insurgencies, as well as the concomitant growth of nuclear arsenals and the compounding horrors of modern war, pacifism lost much of its former vigor as a political and social force in the West. During World War

II, peace advocates came under heavy fire, since many saw them as defeatists or dangerous visionaries who weakened the democracies' resolve as the Axis dictators gained strength, a pattern that continued into the Cold War. That was due partly to the Soviets' heavy-handed use of pacifist symbols and slogans and partly to intense ideological polarity. Pejoratives like crank, Leftist dupe, subversive, and religious fanatic were used to stereotype isolationists, skeptics, and pacifists. The Communist bloc's rejection of "all plans for a world state not originating in the non-Soviet world" fed suspicions in the Western bloc, as did their heavy-handed rhetorical use of "peace," "free," "democracy," and "people's."[31]

The tactical assumption underneath the IAF concept, the idea that air forces would be able to stifle aggression by dropping conventional bombs, if not by merely threatening to do so, weakened as the Cold War intensified. For example, that was Quincy Wright's logic in proclaiming just after World War II that "world order" had "ceased to be primarily technical and has become almost entirely political."[32] As debates within the defense hierarchy over strategic bombing's utility came into public view in the late 1940s,[33] the shortfalls of World War II bombing campaigns were being highlighted in official and popular publications. The U.S. and British Strategic Bombing Surveys presented statistics, maps, and detailed analyses,[34] while Marshall Andrews' *Disaster Through Air Power* argued that the war had been prolonged by allocating men and planes to the bomber offensive rather than build larger ground forces, fleets, and landing craft. Darkling portraits of the bomber offensive in postwar movies like *Command Decision* and *Twelve O'Clock High* contrasted with depictions of strategic bombing as a panacea in wartime films like *Air Force, Bombardier, Winged Victory,* and *Victory Through Air Power*. But those did not dissuade the enthusiasts like the most ardent of air power theorists, Alexander P. de Seversky, who claimed the Air Force held a virtual monopoly in wielding military force in the nuclear age. In a more muted and academic tone, Stefan Possony presented a survey of proposals for international military peace-keeping forces, from the Truce of God to the international air force, to support his case for the United States to impose peace on the world unilaterally by deploying a massive air force sustained by producing ten to fifteen thousand planes a year.[35] The air power proponents clung to the vision of long-range bombers as the only means of delivering strategic nuclear weapons, despite the evolution of rocketry. Until the early 1960s, they imagined some kind of IAF under UN auspices might be formed, along the lines of a "combined international enforcement action" envisioned on the eve of the Korean War by an Air Force general who asserted, "Our total aim will be realized when the obstinate aggressor rulers, faced with the vivid reality of decisive power-in-being, come to alter their policies, discontinue aggression, and sit cooperatively in the United Nations chambers (which) must be equipped with the means of enforcing its decisions and from the technical point of view the easiest means with which to equip it is air power."[36]

The veiled reference to the UN's use of nuclear weapons reflected the Air Force's growing reliance on "nukes," a trend that fueled interservice squabbling that came into public view at the end of the 1940s. After Army's and Navy's careful plans for demobilizing were shattered by the frenzied dismantling of America's military power at the end of World War II, they found themselves competing for scarce funds with the newly created Air Force under the Truman administration's budget stringencies while committed to an increasing menu of missions. The bickering was further aggravated by proposals to unify the armed forces to gain greater efficiency and promote cooperation, a popular idea during World War II that was opposed by the Navy. Admirals Leahy and King, in their final days of influence, brought Admirals Nimitz and Halsey back into line and passed the banner of unification resistance to the relatively young and dynamic James Forrestal. Successor to Frank Knox as secretary of the Navy, Forrestal fought an aggressive delaying action. He commissioned a special study by a senior civil servant, Ferdinand Eberstadt, which grafted the superstructure of the wartime British command system onto the U.S. defense hierarchy under different labels. Eberstadt's report evolved into the National Security Act of 1947, which, like the American Constitution, was designed to limit concentrations of power. In a tragic irony, Forrestal, more exhausted than he or those around him realized, became the first secretary of defense and grew increasingly frustrated as he found himself working against the constraints he helped to devise. At the same time, the U.S. Air Force, created by the 1947 security legislation,[37] was using its monopoly over delivering nuclear weapons to gain dominance in the fight for funds, as well as the contingency, albeit receding, of providing the core of an international air force under United Nations auspices.

Even though opinion polls showed overwhelming public support for an IPF, provided it would be no larger than the U.S. armed forces, the Navy's strong allies in Congress, most notably Carl Vinson, chair of the House Armed Services Committee, opposed such schemes. As the ensuing struggle was paced by mounting international tensions, and the forming of NATO, the superpowers continued to pay lip service to creating some sort of United Nations armed force. The main point of difference between the United States and the USSR, the size of contingents, had appeared in debates over arming the League of Nations: the Soviets preferred equality, while the United States wanted contingents proportional to each nation's population. Critics like political theorist Hans Morgenthau pointed out that the international system and the internal structure of nations were not analogous, and their varying size and power and the existence of empires precluded equity. The empires, however, were fading away, and devolution seemed to be proceeding somewhat faster than many expected just after World War II. As Britain exited India and Burma, the United States left the Philippines, and the Dutch withdrew from the East Indies. But over time imperial devolution still proved to be slow and costly. The British, French, Portuguese, and Soviets tried to retain some of their holdings at a high price in blood and treasure, while the United States was drawn into conflicts in other nations' former imperial possessions in Korea, Vietnam, and the Persian Gulf, as well as its own in the Philippines. In nearly all such "limited wars" of the last half of the twentieth century, air power was used by Western industrial nations to hold on to colonies or to support successor or client regimes against insurgents, Communist and otherwise. In most "limited wars," air power was used in a modified military mode under doctrines and methods not designed for such tasks, partly due to the major Western powers' having dismantled most of their wartime military forces very quickly. As in the 1920s and 1930s, aircraft in those struggles came to symbolize haves battling have-nots. That led United World Federalist vice president Vernon Nash to define "Policing" as a "Euphemism for Empire."[38]

The view ahead was very fuzzy in 1946, when Carl Spaatz, the last USAAF and first U.S. Air Force chief of staff, made a gentlemen's agreement with General Dwight Eisenhower, then Army chief of staff, to consign a major portion of the new Air Force's resources to the Army's needs. That accord was designed to end the feud between the Army's ground force officers and the airmen and to keep the former from joining the Navy in opposing a separate air force. The deal crumbled in 1947, after Spaatz and Eisenhower retired and nuclear weapons shifted the balance of power among the armed services. But the breaking of the compact surprised no one who read Spaatz's article in *Collier*'s in late 1945 on "Air Power in the Atomic Age," in which he staked a claim on a sizable portion of the postwar defense budget.[39] Soon after newsreels of Operation CROSSROADS, the atomic bomb tests at Bikini atoll, gave the world a sense of the scale of nuclear destruction, the new Air Force Chief of Staff, General Hoyt Vandenberg, began shifting USAF priorities to strategic nuclear weapons delivery at the expense of other functions, including support of the Army ground forces in combat. The Navy took the major symbolic hit at Bikini, since warships were the most visible targets, but the deepening Cold War also gave impetus to Vandenberg's efforts. Much of the increased American defense funding that followed the 1948 Communist coup d'etat in Czechoslovakia went to the Air Force, but further impetus came from Mao's victory in China and his alliance with the USSR in early 1949, followed by the Soviets' first test of an atomic bomb. The chairman of the Joint Chiefs of Staff, senior Army general Omar Bradley, declared amphibious invasions obsolete, while media images of Hiroshima, Nagasaki, and Bikini continued to show air power delivering nuclear weapons, and though the Berlin airlift focused attention on the Air Force's transport aircraft, the eagle's share of prestige and dollars was flowing to the newly created Strategic Air Command (SAC).

On a more basic level, the Air Force received major political support from Secretary of Defense Louis Johnson, an air power enthusiast who succeeded Forrestal after the latter, enmeshed in fatigue and melancholia, committed suicide. Beyond that was strong media support, as well as many powerful champions, including Stuart Symington, first secretary of the Air Force, from President Truman's home state of Missouri. At the same time, Bernard Brodie and other academic defense consultants working for the Air Force provided both the tools and logistics for establishing America's strategic nuclear deterrent forces. The sense that Forrestal was a victim of armed service infighting slowed the squabbling only slightly, as SAC became the mainstay of unilateral American nuclear deterrence. Bomber Command, SAC's predecessor organization in the U.S. Air Force, was linked to an IAF role when it was created in 1947, not only in Air Force contingency plans but also in the hopes of its initial commander, General George Kenney, MacArthur's air chief in the Southwest Pacific, 1943–1945, of being the first commander of a United Nations air contingent.

As the Air Force's monopoly over delivering nuclear weapons gave it a budgetary edge over the Army and Navy, the former tried to regain some ground by devising a tactical doctrine based on the use of low-yield nuclear weapons on the battlefield, including an "atomic cannon," small rockets, and the Pentomic Division, which was designed for dispersal and maneuver on the "nuclear battlefield" in central Europe. The Navy came back from a pit of despair, in the depths of which some naval theorists argued that no civilized nation should wield atomic bombs. However, naval aviators staked a claim on a piece of the nuclear deterrent role by devising a plan for launching medium jet bombers against the USSR from aircraft carriers stationed in the oceans around the Eurasian "rimlands,"[40] That concept, based on the relatively limited range of American strategic heavy bombers then in service, the B–29 and B–50, filled in gaps in the arc of Air Force attacks on the USSR in U.S. war plans. In 1949, however, the first true intercontinental bomber was delivered to Strategic Air Command units, the giant B–36 *Peacemaker*, which could fly to Russia and back over the pole without refueling—if it survived Soviet air defenses. When a group of naval aviators challenged the B-36's effectiveness, it triggered a major interservice clash nicknamed the "Battle of the Potomac." Its immediate cause was Defense Secretary

Johnson's breaking of Forrestal's promise to the Navy that it could build a new supercarrier. That bribe, made by Forrestal at a Joint Chiefs of Staff meeting at Key West in 1948 in hope of ending interservice squabbling, failed to end the infighting. After the secretary of the Navy resigned in protest of Johnson's action, the latter relieved the chief of naval operations, which led some angry naval airmen to launch a major publicity and lobbying campaign against strategic bombing, including leaking classified data on the B-36's deficiencies. That "Revolt of the Admirals" culminated in highly publicized congressional hearings in the shadow of ominous developments in the Cold War—the first Soviet atomic bomb test, the Berlin airlift, Chinese-Russian rapprochement, and the forming of the NATO alliance. Nevertheless, much venom had been exchanged in public by the time that General Eisenhower, testifying before Congress, urged restraint, pointing out that the sand box fight offered aid and comfort to America's adversaries and brought no credit to anyone involved.

Whatever prospects the IPF and IAF concepts had at that point went further adrift in all that turmoil. Although a policy analyst asserted at the time that "the domestic origin of conflicts is no longer a barrier to intervention if actual or potential repercussions beyond the borders involve a threat to international peace,"[41] the deepening Cold War put such contingencies in abeyance as Soviet harangues, vetoes, and a boycott crippled the UN Security Council. The outbreak of the Korean War in late June 1950, however, produced a dramatic change by bringing the UN's peace enforcement role and the Air Force's IAF role back to center stage. Soon after North Korea invaded South Korea, the Soviets' absence from the General Assembly allowed Secretary General Trygve Lie to gain approval of armed intervention against North Korean aggression. At the same time, President Truman described American intervention in Korea as participation in a United Nations "police action, rather than a war."[42] Although that assertion was derided by both the Communist bloc and American troops in Korea alike, many internationalists took heart at seeing a group of nations outside a traditional alliance structure take arms against aggression under the banner of an international organization established to maintain the peace. Although claims by Truman and Senator Paul Douglas of Illinois—that UN military operations were an "exercise of police power under international sanction"[43]—dovetailed with earlier attempts to define international forces using force against aggression as something other than war,[44] their assertions proved very flimsy in the rough-and-tumble world of power politics. Nevertheless, Trygve Lie also managed to make some bricks without straw when he gained UN General Assembly assent for his "Uniting for Peace" Resolution in the fall of 1950. While that seemed like a major step toward institutionalizing UN forces, it reached a dead end, and the expectations of American and

United Nations elites proved less realistic than those of academics on the sidelines who pointed out that "The hope that . . . (a) world constitution . . . (imposed) by some kind of international police . . . rests on a misconception of the essence of law and force . . . permanent peace cannot come until it is sanctioned by a sufficiently large section of mankind."[45]

That was not apparent, however, in the opening phase of the Korean War, as strategic bombing to thwart aggression became the centerpiece of American response to the North Korean attack. When President Truman sought counsel from his military advisors in the opening hours of the war, the Army and Navy service chiefs were cautious about intervening, but the senior airmen were as optimistic about winning a significant success through heavy bombing as their predecessors had been a decade earlier. As a result, the initial British and American carrier strikes and tactical sorties from Japan were followed by a massive air offensive carried out by USAF heavy bombers. With the USSR's boycott of the UN allowing the United States to rally support for intervention unhindered by veto or opposition, the General Assembly's resolution sanctioning intervention led some twenty UN members to deploy contingents, spearheaded initially by American and British Commonwealth forces, to support shattered South Korean forces crowding into the "Pusan Perimeter." That, however, was more symbol than substance. Beyond NATO and neutral countries' forces and the British Commonwealth Division, most ground fighting was done by the refurbished South Korean Army and by American soldiers and Marines. The U.S. Air Force bore the brunt of the air war with support from Navy, Marine Corps, and a few British Commonwealth elements. Despite some attempts to cast the initial bombing campaign in that light, the overall effort fell well short of being an international air force. And contrary to Secretary General Lie's hopes, the United Nations Command under General MacArthur proved to be only a figurehead, as the latter blocked attempts to create an effective coordinating structure and issued reports to the Japanese media before sending them to UN headquarters in New York.[46]

As UN forces struggled to retain a foothold in southeastern Korea from June to mid-September 1950, the U.S. Far Eastern Air Force's Bomber Command's B–29s and B–50s flew from bases in Japan and the western Pacific and attempted to sever the North Korean People's Army's lines of communication around the Pusan perimeter, while hammering their forces massing for assaults. By mid-September, Far East Bomber Command (FEBC) had "serviced" all major targets, and the vast number of civilian deaths triggered massive condemnation from the Communist bloc, some neutral countries, and NATO Allies as well.[47] At the same time, however, it fed some enthusiasm in the United States for further use of strategic bombing as the "biceps and righteous fist to the United Nations."[48] In mid-September, a coordinated breakout by UN ground forces from the Pusan Perimeter and an amphibious landing at Inchon on the west coast of Korea shattered and routed the the North Korean People's Army (NKPA). Those gains were all erased two months later when Chinese Communist forces crossed the Yalu River and drove UN ground forces south of the 38th parallel, the prewar boundary between North and South Korea. As static ground warfare raged in central Korea for the rest of the war, UN air elements, mainly American, maintained air dominance over most of the peninsula, while engaging the Chinese and North Korean air forces in jet–to–jet battles over "Mig Alley" in northwest Korea, close to the PRC's air bases and far from those of the UN forces. Ironically, as Soviet support made China's air force the fourth largest in the world, it also became a kind of Communist-bloc international air force. U.S. airmen reported sighting Caucasian Mig pilots, but the scale of involvement of "volunteers" from eastern bloc nations and the Soviet Union doing their "internationalist Socialist duty" only became apparent a generation later.

Contrary to American air commanders' hopes at the outset, the sustained strategic and tactical air operations throughout the Korean War failed to keep China and North Korea from deploying large armies or launching major offensives.[49] As the conflict dragged on until the armistice of July 1953, the image of the USAF as the UN's air arm faded away along with the prospect of creating standing UN military forces and a command structure.[50] American airmen found increasingly tight rules of engagement all the more frustrating since minimal constraints had allowed them to design the opening air offensive along the lines of World War II, using radar bombing and incendiaries to convey the "psychological impact of bringing war to the people." The contradictory logic of imposing punishment on the masses in an especially oppressive dictatorship notwithstanding, the airmen's exasperation became dramatically apparent in early 1951 when FEBC commander General Emmett "Rosie" O'Donnell proposed using nuclear weapons to lay the "lash" on China.[51] That demand, and his feisty testimony before Congress aggravated some NATO members' fears that the Korean War might inadvertently lead one or the other superpower—or both—across the nuclear trip wire, either by accident or by intent.[52] From the outset, Western European leaders were concerned that America was being distracted from commitments in Europe, by the Far East as it had been during the first year of involvement in World War II. President Truman tried to allay those anxieties by dispatching forces to Europe equal in size to those deployed in Korea. As the Korean War progressed, NATO defenses were also being bolstered by the deployment of tactical atomic weapons, substantially reducing the need for large armies envisioned in original plans for forming the North Atlantic alliance. Exclusive American control of nearly all the "nukes" in the non-Communist world and the development of nuclear weapons by Britain and Russia further weakened both the IPF and IAF concepts, as did the failure of heavy strategic bombing to deter aggression in Korea before the war, to win a clear decision after it began, or to prevent large Chinese Communist armies from attacking. After the Korean War, the sense in some quarters that the use of tactical atomic weapons might have ended the conflict quickly obscured the potential diplomatic and political costs. With emotions running high in the councils of power as the Cold War intensified, many among the elites seemed ready to accept those risks, come what may. Even a neo-Wilsonian felt compelled to proclaim: "Better the physical destruction of humanity by the atom bomb than the elimination of morals from the life of man."[53]

Against that darkening landscape, whether as a consequence or not, the war in Korea ended as the Eisenhower administration was threatening to use tactical nuclear weapons there if the North Koreans rejected an armistice. At the same time, Secretary of State John Foster Dulles was articulating his policy of "massive retaliation," that is, extending the American nuclear umbrella over dozens of nations involved in mutual security pacts with the United States as part of a broadening of Truman's doctrine of containing Communism. At the same time, the administration's "New Look" defense policy shunted more of the defense budget toward the Air Force and the Navy's strategic deterrent forces. Although the Korean conflict would be described a decade later as having "soured the prospects for quasi-collective security actions,"[54] the outlook was dimming well before that. Walter Lippmann's 1950 declaration that collective security was akin to "burning down the barn . . . to roast a pig" and "too crude, too expensive and too unreliable for general and regular use" was seconded by an analyst who rejected collective security as a "fallacy and a fantasy" and deemed the idea of an IPF "an empty dream."[55] It was not surprising then that British attempts to form a European air force within the framework of the European Defence Community in the early 1950s came to naught. Korea also highlighted both the weakness of analogies between military forces keeping the peace and police forces fighting criminals and the imperialistic flavor of using air power to coerce. But despite widespread criticism of American strategic bombing in that "limited war" and American aversion to becoming entangled in another limited war, some U.S. airmen became interested in the Royal Air Force's "air control" in Iraq and Aden in the 1920s and 1930s as a model for "unconventional warfare." A major U.S. Air Force research program, Project CONTROL, launched in 1954 to review those experiences, eventually produced twenty-one volumes of data. But, as an official historian later concluded, it failed to "fully demonstrate . . . that the project thesis"—that such methods might be used by the United

States Air Force in the Cold War—"was completely applicable to continuing national security problems." At the same time, Eisenhower's secretary of the Air Force, Thomas Finletter, opposed expanding United Nations disarmament inspection and enforcement authority on the grounds that it would be an "affront to national pride and independence," or at least those of the superpowers.[56]

Along with many nuclear disarmament schemes that dead-ended during the Cold War,[57] a myriad of attempts to structure international forces came to naught, including Senator Henry Cabot Lodge's plan to recruit refugees from Soviet-controlled Eastern Europe into an American foreign legion, the Pleven Plan for forming a "supranational" European Army, and the multinational force proposal for placing crew members from all NATO nations on a ship bearing nuclear weapons that would cruise the high seas on a random course. Such schemes all foundered to varying degrees on the old dilemma of national sovereignty, along with the paradox identified in the mid–1950s by the dean of American defense scientists, Vannevar Bush: the need "to be simultaneously ready for the all-out war and also be ready to handle the secondary wars by modern means that are just the means of World War II dressed up a bit."[58] That quandary was highlighted in the Far East less than a year after the Korean War when the French, moving forward independently to develop nuclear weapons but having none in hand, faced defeat in Indochina, where the cream of France's Expeditionary Force, some 10,000 troops, were surrounded at ten-to-one odds by the Viet Minh at Dienbienphu, a remote stronghold in Laos. The crisis brought a deputation of French officers to Washington, D.C., to lobby for a major American air strike to break the siege. France had already received several billion dollars worth of American military aid to their forces in Indochina and Truman's secret assurance that the United States would not sign a separate peace in Korea while war was raging in Indochina. Eisenhower, who had not honored that line of credit, now sought counsel in many directions. Admiral Arthur Radford, chairman of the U.S. Joint Chiefs of Staff was eager to act and ordered French insignia painted on American B-29 bombers and carrier planes in the Far East. Under the code name Operation VULTURE, some steps were taken in anticipation of employing tactical nuclear weapons. While Vice President Richard Nixon proposed dispatching 250,000 American troops to Indochina, other key players urged restraint, including Senate Majority Leader Lyndon Johnson, Secretary of State John Foster Dulles, and Prime Minister Winston Churchill. Eisenhower chose the middle way, and the French garrison surrendered on May 7 as their last outposts were overrun.

The Dienbienphu affair underscored the flimsiness of the NATO alliance in "out-of–area" contingencies as well as the durability of the view that air power could bring quick and decisive force to bear in limited war, even though it had not done so up to that point. The French delegation's entreaties were based on that assumption, and it went unchallenged, as the issue was seen as one of military but not diplomatic and political relevance. Some details were left out of focus, like the Viet Minh besiegers' trenches being crowded so closely around the French positions, that using heavy conventional bombs or nuclear weapons would have inflicted major casualties on the defenders. Beyond that, any tactical advantage gained by such a notional attack also would have yielded a substantial psychological warfare debit. The Communist bloc made much propaganda hay out of American bombers killing many thousands of civilians in Korea and, along with others on the Left and pacifists, used Hiroshima as a symbol of techno-bullying and racism. The appearance of unaligned neutral nations from the late 1940s onward, most of them former colonies, also placed constraints on the major Western powers' using weapons of mass destruction, conventional and nuclear, in limited wars and insurgencies. During the Dienbienphu crisis, even so ardent an imperialist as Churchill opposed helping France retain its colony under the guise of resisting Communism.[59]

Between the Korean and Vietnam Wars, such blurrings of motive increased as the United States extended its web of unilateral and multilateral defense arrangements, while Britain and France reduced their overseas involvement. The creation of UN standing military forces became an ever slimmer prospect as "limited" wars flickered across the Eurasian "rimlands," in keeping with Nicholas Spykman's prediction in 1943.[60] A very faint flavor of an IAF was visible in Anglo–American nuclear war plans in the 1950s and 1960s, which assigned RAF V–Bombers the task of blasting a corridor through Soviet bloc defenses in Central and Eastern Europe, over which the U.S. Air Force's Strategic Air Command would fly on the way to the Soviet heartland. That was a bilateral arrangement, and an exclusive one. As uneven as the Anglo-American "special relationship" was, the French deemed it exclusionary and threatening when they created their own nuclear deterrent in the form of the *force de frappe* (strike force) of *Mirage* jet bombers and, later, a strategic nuclear submarine force. Relations between the major Communist powers were even less harmonious. While Soviet Premier Nikita Khrushchev challenged the West by coining the rubric "Wars of National Liberation" in 1956 to placate his Chinese comrades who urged the USSR to join them in confronting the capitalist bloc more aggressively, that attempt failed to prevent the rupture in Chinese-Soviet relations half a decade later.

Throughout the 1950s, the role of "the United States Air Force" as "the main air bulwark of the United Nations" continued to fade against the backdrop of Cold War and was further undermined by the Eisenhower administration's "massive retaliation" policy and its growing network of unilateral, regional defense treaties.[61] Dulles, however, soon recognized that massive retaliation, deemed "brinkmanship" by his critics, was a brief passing phase in the Cold War, since it depended on the United States to maintain a technical edge in the ongoing rivalry with the USSR that included expensive bomber, rocket, and air defense programs and related communications systems. It was also becoming obvious that nuclear weapons could not be fine-tuned for use in small conflicts and that massive destruction and horror might result from detonating even a single weapon in anger, or from even a minor clash between superpower conventional military forces. Nuclear theorists became increasingly concerned about both maintaining "strategic stability" and the growing uncertainties about the concept of deterrence, which was wrapped around an inner core of literal lunacy. Some hidden costs were revealed by the "Lucky Dragon" incident in the mid–1950s, when a Japanese fishing boat was covered by radioactive fallout from an unexpectedly powerful American hydrogen bomb test in the Pacific. In the aftermath, scientists learned that the atmospheric nuclear tests under way since 1946 had contaminated the upper atmosphere in the northern hemisphere far more than experts had estimated and were producing a long–term fallout separate from that produced in the immediate zone around nuclear explosions. Their mutual alarm led the United States, the USSR, and Britain to sign the Atmospheric Test Ban Treaty of 1963 and fed rising public anxiety about the nuclear arms race in the Western democracies from the mid-1950s on. This anxiety was also fanned by expanding civil defense programs and such films as *Dr. Strangelove: Or How I Learned to Stop Worrying and Love the Bomb, Fail-Safe, On the Beach,* and *The War Game.* The Cuban Missile Crisis of 1962 triggered a widespread panic and highlighted the paradox that Bernard Brodie identified when he framed American deterrence theory in the late 1940s, that "nukes" had no rational utility beyond serving as a threat or as a form of retaliation. There was, by implication, no reasonable rationale for "first use."

Despite those mounting concerns, no substantial antinuclear movement emerged in the United States during those years, due, in varying degrees, to the decline of pacifism, the soothing effects of diplomatic efforts that produced the Test Ban Treaty and subsequent agreements like SALT I and II, and some military professionals and defense analysts' open skepticism regarding the utility of nuclear weapons. In 1959, retired Army Chief of Staff General Maxwell D. Taylor set forth a new "spectrum of response" defense doctrine in his book *The Uncertain Trumpet,* proposing that the United States should back away from reliance on "massive retaliation" and configure its armed forces to deal proportionately with conflicts on a scale from low-level insurgency to a "general war." In the early 1960s, Taylor became national security advisor in the Kennedy administration, and later chaired the Joint Chiefs of Staff, pushing the "spectrum of response" and counterinsurgency doctrine—COIN—to the forefront in American defense policy.[62]

After the "Battle of the Potomac," interservice friction eased slightly when the Army and Navy received larger shares of the nuclear arsenal, although the Air Force, Army, and Marine Corps continued to wrangle over who would control close air support. One of the most substantial effects of Taylor's "spectrum of response" was the increase in "special operations," including the creation of new elite units under the "black," or secret, budget and assigning tactical and low–intensity supporting roles to both Army and Air Force covert elements.[63] On the eve of the Vietnam War, the IPF/IAF concepts lay on the far margins of defense policy and analysis, partly due to declining Western European involvement in world affairs and partly due to increasing American power and unilateralism. While there were more and more nuclear weapons, there were also more and more constraints on their use, as well as on conventional weapons that might trigger the use of "nukes." During the decade between the Korean and Vietnam wars, the changing matrix of power was symbolized by wars of imperial devolution in places like Malaya, French Indochina, Kenya, and by the Bandung Conference of "Third World" nations in 1955, which excluded superpowers and former imperial overlords. That gathering of Third World leaders highlighted the growing leverage exercised on the teeter-totter of superpower rivalry by nations formerly under European imperial rule. "Gunboat diplomacy" and "air policing" were now linked with the old colonial order as Western pundits, government officials, and military professionals grappled with the intricacies of what Morton Halperin labeled "limited war in the nuclear age."

As COIN became a trendy acronym in American defense circles in the late 1950s, analysts found it increasingly hard to discern patterns and devise stratagems, since each crisis point offered a different mix of intricate factors like imperialism, economics, ethnic tensions, regional politics, strategy, diplomacy, and the putative influence of Communist subversion. The spectrum–of–conflict paradigm posed a special dilemma since, in "unconventional warfare," subtle and faint aspects like propaganda, intelligence, or informal elite relationships often proved to be more crucial than military high technology and firepower. Reliance on military firepower might yield more debits than credits, especially strategic bombing, which had once seemed to offer the brightest promise for multinational peace enforcement. As western analysts searched for patterns and nostrums in that maze, they framed mechanistic and simple metaphors like fire–fighting—a glass of water might easily extinguish it at the outset, but it might quickly become a conflagration if left unattended. A major problem lay in matching the size of the glass, bucket, or hose to the size of the blaze.

Western industrial powers involved in "brushfire wars" usually found themselves ensnared in webs of propaganda woven of strands of Communist rhetoric, nationalism, and anticolonialism. What one observer might view as an attempt to maintain order and security, another might see as oppression and exploitation—a problem that had bedeviled proponents of an IPF since Paris in 1919. Such metaphors and other theories, including Che Guevara's focoism, led analysts and policymakers to see COIN as more generic than it proved to be in various hot spots during the Cold War. Some Western Cold Warriors who framed COIN theories, like Edward Geary Lansdale, Sir Robert Thompson, and Roger Trinquier, saw insurgencies as complex blends of such ephemeral and idiosyncratic dimensions as ideology, propaganda, and culture, but others saw them in simpler, bolder terms—as part of a Soviet design to dominate the world, as "bad guys" trying to unseat a properly constituted authoritarian regime, or as a military problem to be solved by military means.

The one clear pattern amid all that complexity was that there was no clear pattern. Prior to the Vietnam War, various configurations of air power yielded a clear tactical advantage in some smaller and shorter "limited wars" like Kenya, Algeria, and the Philippines, but not in bigger and longer ones—Korea, Greece, Indochina, and the Chinese Civil War. The pattern was also mixed in respect to the bigger and shorter conflicts—the Indo-Pakistani Wars and the Arab–Israeli clashes of 1948 and 1956. The continued bubbling up of such hot spots conformed with John Burton's warning that peace could not be enforced and with critiques of the League to Enforce Peace's proposals that had appeared half a century earlier. Dismissing international law and police analogies as fallacious, Burton declared "international enforcement . . . a tyranny [that] as such will never be accepted by states against which it is exercised." Flipping a popular perception on its back, he argued that "policies and structures traditionally employed to prevent the emergence of conflict" were "the greatest potential source of open warfare."[64]

In considering that tangle of events and concepts, it is not surprising that American policy and strategy, and policy analysis as well, from Korea to Vietnam, and arguably well beyond, tended to be ambiguous if not schizoidal. At the same time that proposals for IPF/IAF schemes continued to appear in Congress, along with U.S. initiatives in the UN, American regional security treaties and pacts weakened the fiber of world organization and reduced the likelihood that any kind of multinational international policing entity would be formed.[65] Despite the strong leitmotif of unilateralism in its defense policies and diplomacy, the Eisenhower administration expressed interest in establishing a UN constabulary throughout the 1950s, even though the concept of collective security was being left well off to the side in U.S. defense circles as planners grappled with such complex problems as orchestrating the expansion of the TRIAD, arms control negotiations, technical advances, interservice friction over "roles and missions," and uncertainties and intricacies of COIN and limited war. Nevertheless, despite those intricate cross-currents and the very thin prospects for creating a standing UN peacekeeping force, a few tireless enthusiasts continued to propose IPF/IAF models, hoping against hope that if the United Nations raised a small reliable multinational volunteer unit like the French Foreign Legion, it would provide the kernel of ever-larger standing UN forces that might ultimately evolve into a major deterrent to aggression around the globe.[66] Architects of such models, like their predecessors between the world wars, were not daunted by the fact that the superpowers would not allow it or the lack of support from a major political constituency, let alone the general public.

In 1956, the Suez affair provided another demonstration of the limits of air power. Those holding the upper hand in the air—the British, French, and Israelis—failed to win a clear strategic victory over Egypt, mainly because the United States and USSR both censured them in the UN and forced a halt to the fighting. At the same time, Suez spotlighted the issue of standing UN IPF units when a United Nations Emergency Force (UNEF) of six thousand was deployed along the Israeli-Egyptian frontier after the ceasefire. That raised questions much like those heard in the hallways of the League of Nations: What was the legal status of such forces? What precedent was the deployment setting? Where might major powers' involvement lead if UN military operations touched on their interests?[67] When the Eisenhower administration engaged in "nuclear diplomacy" in the final phase of the Korean War by brandishing nuclear weapons to force an armistice, the United States had a significant edge in nuclear weapons over the USSR and was involved in a conflict alongside troops from other nations serving under the UN flag. Historians have debated whether moving an atomic cannon across the Pacific in stages actually ended that war, but they have paid less attention to the question of where using "nukes" under the UN's auspices, however tenuous its sway, might have led. In any event, both American nuclear dominance and UN prestige were both declining as Soviet nuclear forces grew during the 1950s, and China developed an atomic bomb.

Paradoxically, concern in some quarters in the United States about the UN's expanding authority peaked in the early 1960s, when Secretary General Dag Hammerskjold deployed UN contingents in the Congo amid the major upheavals that followed Belgium's granting its independence. Media reports of colonial successor states and Cold War neutrals' troops committing atrocities, as well as concern that contingent commanders were following national and personal agendas beclouded the first deployment of combat troops under United Nations command and control.[68] In the United States, where the "civil rights revolution" was under way, images of Indian and African troops fighting under the UN flag heightened segregationists' fears that such outside forces might intrude into their local matters. Those anxieties led an advocate of a stronger United Nations to deny that "the Congo [was] a precedent for the UN going into Mississippi" since "by no stretch of the imagination" was "the situation in Mississippi . . . a threat to international peace."[69] Such anxiety about *Quis custodiet custodies ipsos* (Who watches the watchmen?) was, however, not confined to rustic regions of the Great Republic.

In that perplexing context, American public attitudes on the IPF issue were far more diverse and contradictory than they had been during World War II. Bumper stickers calling for U.S. withdrawal from the United Nations appeared throughout the United States in the early 1960s, and opponents of "one-world government" invoked logics ranging from a roseate view of America as an exceptionalist Arcadia to citing the Bible to support their arguments. The former stance was articulated by Colonel Robert McCormick, publisher of the isolationist *Chicago Tribune*, who saw "the glory of this country" in its being "separated . . . from the strife of the old world" so that "men of all races and countries have learned to live together in peace." "If we are taken into a world league," he warned, "we shall never be quite the same as we have been. The corrosive agencies of Europe will have set to work on the American amalgam."[70] An example of the latter was Southern fundamentalists' view of racial and cultural differentiation as God's handiwork and anticipation that "the blending of worldwide governments, ethnic groups and religions would signal the coming of the Antichrist."[71]

Opinions also varied widely in the realm of partisan politics, where the IPF concept continued to receive support from such diverse sources as liberal Democrats like Henry Wallace, and moderate Republicans Wendell Willkie and Henry Cabot Lodge Jr.[72] The latter proposed recruiting an American equivalent of the foreign legion from European refugees who had fled Communist countries. One advocate of a stronger international order saw a brightening prospect in the fact that the UN had actually raised "a small international police force" which might become large enough "to protect smaller nations from each other and to disarm them if necessary . . . long before there is one able to impose its will on the super-powers."[73] Another enthusiast expressed hope that it might lead to a subtle demilitarization of the traditional role of the soldier under the hypothesis that "an army is entirely too dangerous an institution to be embroiled in controversy and to be characterized by strong loyalties and identifications."[74] Closer to the center of the political spectrum, another analyst, after surveying the evolution of peace theory and organization, concluded that the United Nations would never evolve into a world government and warned that the increasing trend toward armed intervention would "run into all the problems which destroyed the League, and, like the League, it [the UN] will end by being discredited and destroyed."[75] The ripple of outcries against the UN across the United States led some observers to judge "the attempt to develop a workable collective security organization as stymied," especially when schemes for an international police force, however carefully and cautiously crafted they might have been, were linked to concepts like world government and total disarmament.[76]

Despite mounting pessimism about the UN's future and prospects for creating a kind of IPF or IAF, some peace enforcement proponents took a further visionary leap into the realm of outer space. While the Army Air Forces had been interested in rocketry from World War II onward, and the Army, Navy, and Air Force created their own development programs, a major American effort in that area did not come until after the Soviets tested their first intercontinental ballistic missile (ICBM) and launched the geophysical telemetry satellite *Sputnik* in the fall of 1957. In 1960, against the somber backdrop of the chronic failure of early American rocket tests, the editor of *Air Force* magazine blended the moribund IAF concept with the emerging space mission in suggesting the creation of a "spaceborne warning patrol" followed by "the pooling of the world's strategic attack force into an international enforcement agency, geared to the global warning network, directed by the United Nations."[77]

That grandiose, gossamer vision, like other IPF schemes, was soon scattered by rising gales on the international scene. When U.S. Air Force planes carried Belgian paratroops to Stanleyville in the Congo affair, it not only raised protests on the American Right but led a moderate observer to deem the expedition a "challenge [to] the assumption of the infinite repeatability of the UNEF type of the United Nations contribution to peace and security."[78] It was a moot point. Prospects for increased UN armed intervention faded rapidly after Secretary General Dag Hammarskjold died in a plane crash, and the United States was becoming more deeply involved in the growing "brushfire war" in Vietnam. Deceptively small at the outset, that conflict grew in momentum outside the UN's orbit. While Suez showed America's major NATO allies the geographic limits of the alliance, Western European powers' views seemed less and less relevant as the situation in Vietnam heated up rapidly after John F. Kennedy took office in early 1961. America's power was growing dramatically. The first U.S. Navy nuclear powered Polaris submarines carrying strategic missiles were on station, and the Air Force was putting ICBMs and new strategic bombers into service. Even though that "Triad" of nuclear weapons delivery systems weakened both the Air Force's dominance over deterrence and the conventional bomber force that underlay the IAF concept, it seemed possible for a brief moment in the summer of 1961 that an IPF might be created under U.S. and

Soviet authority. The Clark–Sohn plan was suddenly brought center-stage in Cold War diplomacy during arms control chats between Soviet emissary Valerian Zorin and veteran American civil servant John J. McCloy. Both seasoned veterans of bureaucratic infighting and diplomacy, they concocted an alloy of what McCloy's biographer called "toughness and the most utopian form of internationalism." Drawing on the Clark-Sohn plan, their "Joint Statement of Agreed Principles for Disarmament Negotiations" included an agreement to use an international police force to oversee disarmament and ultimately form a United Nations peace force. Despite apparent approval by the Soviet and American governments, it proved no more viable than previous IPF schemes.[79] The most immediate causes of the reversal seemed to be the Berlin Crisis and Vietnam, but disintegrative forces also included the relatively low accuracy of the Navy's strategic missiles and the superpowers' deployment of hydrogen bombs and ICBMS, all very far from using conventional bombs in relatively "surgical"—precision—attacks by an IAF. At a distance, it appears more as a curious anomaly, symptomatic of the angst of those times.

On the eve of the Vietnam War, many observers saw collective security as totally defunct and the United Nations as having evolved from a "mountainous" determination to establish an effective peace-keeping mechanism to "a mousy commitment."[80] Others saw peace enforcement concepts as afflicted by a "Paralysis of Collective Responsibility"[81] and "an international police force" as "an impotent anachronism without an international authority."[82] Some students of political theory and diplomacy attributed that to superpower and regional polarities,[83] echoing 1920s skeptics in their assailing attempts to equate laws within nations with the flimsy sketches offered by architects of global order.[84] Caught in those quickening tides of *Realpolitik*, some serious-minded policy analysts still clung to the fading hope that some sort of IPF might be created.[85] In 1963, for example, as the crisis in Vietnam mounted, pacifist military historian Walter Millis and a colleague proposed creating national police forces to serve as basic building blocks for a multinational "police type of world organization"[86] under a "New World Authority" along the lines of models proposed by Theodore Roosevelt in his 1910 Nobel Prize address, and by collective security enthusiasts during World War II. At the same time, D. W. Bowett pointed out that even though the UN lacked a standing force, armed elements under its auspices had been deployed continuously somewhere in the world since 1948, which he deemed all the more amazing since it was "extraordinary . . . that the United Nations has been able to establish military forces at all."[87]

Unlike the Korean War, the UN was left off-stage as the Johnson administration waded into the swamp of Vietnam. Under considerable American suasion in 1950, the UN had become a very different entity as dozens of new nations joined over the next decade, many of them condemning the United States as the successor or ally of their former colonial overlords, a mentor of authoritarian regimes, and/or racist. However valid such perceptions were, weak coordination among the U.S. armed services and linked organizations and a lack of clear strategy and goals produced a melange of efforts, which generated, all at once, a large ground war, covert operations in the hinterlands, a massive civil assistance program, and Operation ROLLING THUNDER, an extended bombing campaign against North Vietnam, equal to all those of World War II in duration and far surpassing them in scale. Even before the air war began, air power was a major issue in the presidential campaign of 1964. Incumbent Lyndon Johnson struck a posture of pacific moderation while characterizing his opponent, Republican senator Barry Goldwater from Arizona, a major-general in the Air National Guard, as the advocate of fighting a major war in Vietnam. While Goldwater favored a vigorous air campaign, he opposed committing ground troops and suggested cutting losses and withdrawing if bombing failed to yield results. Many, including Democratic Party tacticians, linked those views with the senator's proposal to give NATO field commanders greater latitude in using nuclear weapons.

In August 1964, reports of North Vietnamese attacks on U.S. destroyers in the Gulf of Tonkin allowed Johnson to strike the pose of a cautious commander in chief when he ordered limited retaliatory air strikes. That incident led Congress to pass the Gulf of Tonkin Resolution, which authorized the president to use force to protect American interests. While U.S. warships involved in that incident were supporting covert operations, U.S. diplomats and advisors in Saigon had been urging military action to demonstrate American resolve as North Vietnamese and the Communist-dominated National Liberation Front forces brought increased pressure to bear on the series of shaky regimes that had followed the coup that toppled the Diem regime in October 1963. Johnson used the blank check of the Gulf of Tonkin Resolution sparingly during the election campaign and implied he would steer a cautious course in Vietnam if reelected. Keeping preparations for a major escalation secret, he won a major landslide victory, and launched ROLLING THUNDER in early 1965, along with ordering the the first major increment of U.S. ground combat forces ashore at Da Nang.[88] ROLLING THUNDER was carried out by two major forces, U.S. Navy planes operating from "Yankee Station," aircraft carriers in the Gulf of Tonkin off North Vietnam, and U.S. Air Force elements based in Thailand. Paradoxically, those strategic missions were performed by fighter-bombers designed for tactical operations, while massive ARCLIGHT tactical raids were conducted by B–52 heavy bombers based in the Philippines and on Guam, planes originally built to carry nuclear weapons on long-range strategic missions.

American and South Vietnamese air operations in the Vietnam War and those of other nations involved in the conflict, like Australia, were farther from the IAF model in concept and structure than those on either side in the Korean War. Both conflicts, however, lay in the shadow of the superpowers' nuclear buildup, and in each case, the adversaries adhered to tight rules of engagement. South Vietnamese aircraft rarely entered North Vietnam, and North Vietnamese aircraft launched no attacks on South Vietnam or on Allied ships on the high seas. American aircraft refrained from bombing certain types of targets in ROLLING THUNDER strikes, and Viet Cong terrorists in South Vietnam left U.S. command-and-control facilities alone.[89] Political concerns and constraints were visible as Lyndon Johnson and members of his staff picked targets during the early phase of ROLLING THUNDER, using "escalation theory" to apply force in graduated increments as a "signaling" device. Although it was clear by early 1966 that the air offensive was not yielding clear-cut results, it continued in fits and starts until the spring of 1968. Nor did Administration officials agree regarding goals and results. Was it intended to cut supply lines, show American commitment and resolve, serve as a bargaining token or signaling device, bolster South Vietnamese morale, or force the North Vietnamese to negotiate or capitulate?[90] It did all of those to varying degrees, but measuring the effects was a murky business. Early in the air campaign, Secretary of Defense Robert McNamara was dismayed to learn that 267 sorties had hit less than 20 percent of targeted buildings.[91] By 1967, the lack of clear results led to major clashes between the Joint Chiefs of Staff and McNamara, which a historian later defined as "what might have become the gravest crisis in civil–military relations in modern U.S. history."[92]

As tensions mounted over the Vietnam War in American society from 1965 to 1969, ROLLING THUNDER became a major target of the antiwar movement and generated an intense exegesis over air power's effectiveness among defense intellectuals and academic critics.[93] Although the Korean War led some analysts to conclude that a thinly industrialized, labor intensive system could not be effectively impaired by air attack, that had also been true in World War II in Europe, where targets were technically complex and concentrated.[94] Some critics of the Vietnam air war, like the Air War Study Group at Cornell University, brought some scientific rigor to their briefs,[95] but others did not. A journalist, for example, asserted that "well-known air force studies of the strategic bombing of Germany . . . tend to prove it is the most wildly inefficient form of mass destruction yet devised,"[96] even though the Strategic Bombing Surveys (SBS) were not conducted by the Air Force, and offered no such conclusion. However, they did point out a need for careful planning and provided some guidelines for waging air warfare more effectively. (Extracts of those guidelines were used in the mid-1950s for civil defense planning.) One U.S. SBS conclusion relevant to Vietnam was that while air raids might turn those being bombed against their governments, they might also generate hostility toward the attackers,[97] a paradox visible since the First World War.[98]

By the Vietnam War's end in 1975, over six million tons of aerial munitions had been "delivered" by American and South Vietnamese forces.[99] Once again, air dominance failed to produce a clear resolution, and as in earlier cases, tight rules of engagement led airmen to offer an extenuating plea much like those heard during and after World War II and in the Korean War—"if only air power had been applied all-out."[100] And like those instances, the Vietnam air offensive left the utility and effect of coercive bombing out of focus. It was not clear, for example, how much frequent bombing pauses allowed North Vietnam to bolster its defenses or delayed its coming to the conference table, or to what extent air attacks drove the civilian population closer to the Hanoi regime or enhanced its power.[101] Critics of the Johnson administration's Vietnam policy abroad and in the United States played up the contradiction between America claiming to be the champion of democracy and using such weapons of mass destruction as B–52 bombers, napalm, and defoliants.[102]

Amid those uncertainties and contradictions, confidence in the utility of air power and in the idea of an IPF both maintained momentum. In 1969, for example, as American forces in Vietnam peaked at 650,000 in the aftermath of the failed air offensive, Quincy Wright, the much respected scholar of war dynamics, included a design for a denuclearized "world peace force" when he updated his classic *A Study of War*. A variation on an old theme, like his suggestion for using artful propaganda to shape world opinion in support of a "universal world state," it echoed some peace-through-coercion schemes designed early in the century.[103] At the same time, other defense analysts pondered whether a more sophisticated coercive bombing scheme might not have been more successful, although such attempts to draw "lessons" from history looked more like lawyers' searching for evidence to support a particular position.[104] Like the works of Marx and Hitler, the Bombing Surveys were rarely read by those who cited them in the debates over air power's efficacy. Other critiques of strategic bombing reflected armed service, pacifist, or humanitarian biases or concentrated on mechanistic aspects like cost effectiveness. Some of President Johnson's advisors had been involved in the Strategic Bombing Survey, most notably Walt Whitman Rostow, who proposed attacking the North Vietnamese power grid to no avail.[105] After Richard Nixon succeeded Johnson, the most bitter pill was set on the plate—the conclusion of intelligence analysts that ROLLING THUNDER had stimulated North Vietnamese military recruiting.[106] Three decades later, when much of the dust had settled, McNamara closed a very large circle in asserting that "the United Nations Charter offers a far more appropriate framework for international relations in the future than from the doctrine of power politics."[107]

The last year of direct American military involvement in Vietnam, 1972, was marked by the extensive use of precision guided missiles (PGMs), or "smart bombs," during the last two air offensives against North Vietnam—Operations LINEBACKER I and II. Although two previous commanders of the 7th Air Force rejected the urging of the secretary of the Air Force that they use PGMs, their successor, General John Lavelle, was of a different mind. He had worked on a special PGM project and, as senior U.S. airman in Vietnam, employed them in LINEBACKER I in the spring of 1972 to break up a major North Vietnamese armored force massing to invade South Vietnam. Having exceeded the rules of engagement, he was relieved from command.[108] LINEBACKER II, launched at Christmastime 1972 after the North Vietnamese had walked out of the Paris Truce Talks, saw a renewed use of PGMs and the only deployment of B–52s against North Vietnam in that fourteen-year war. Three weeks later, the Paris conference reconvened, and a truce was signed that ended American military involvement in Vietnam under the rubric "peace with honor."[109]

When the war ended in 1975, the IAF concept remained well out of focus in public policy processes and discourse. Despite the blurred patterns of the air war in Vietnam, a few enthusiasts still urged the creation of some sort of IPF. Once again, the ever resolute Grenville Clark sounded a call for a "World Police Force," offering a version of his earlier design much like the one presented by Herriot at Geneva half a century earlier and the original Spanish initiatives both in its detail and impracticality.[110] Such schemes weighed lightly in the concerns of President Richard Nixon and Henry Kissinger, his national security advisor and later secretary of state, as they struggled to restore the international balance of power in America's favor through traditional diplomacy, that is, bilateral, multilateral, and regional agreements. In the early 1970s, many feared and others hoped that the Vietnam debacle had ended significant American involvement in world affairs. Polls showing strong public support for disengagement as the spectacle of the world's richest and most powerful nation being reduced to a "pitiful giant" puzzled and frustrated many Americans, although that was not a unique experience for major nations in the history of power politics.[111] A generation earlier, critics of both the IPF concept and the "League with teeth" had recognized that possibility, but the Korean and Vietnam Wars' demonstrations of the limited capacity of great powers to impose their will nor contrary opinions did not daunt Nixon and Kissinger in their attempt to rebuild American power and credibility. In consonance with those trends, one analyst, after tracing broad patterns of universalist failures and regional successes in international affairs, defined collective security as a "high academic" construct of no utility in a world in which impulses to maximize self-interest and to make deals would prevail over altruism.[112] Some saw Vietnam as an anomaly and expected future American administrations to avoid such pitfalls as they continued to make and honor commitments and intervene in world affairs. Others suggested that domestic support might be secured by using more adroit methods than Johnson's and saw strategic bombing as a "militarily useful and militarily sound means of achieving important objectives."[113]

As the "post-Vietnam malaise" faded slowly throughout the 1970s, during the Ford and Carter administrations, the U.S. armed forces passed through a "hollow" period of lean funding while wrestling with major changes in technology. As "Aerospace Power" had evolved throughout the 1960s, and ICBMs and the Navy's Polaris submarines displaced bombers as the primary nuclear weapons delivery systems, three major strategic bomber types—the B–36, B–47, and B–58—entered and left service without being used in combat. Weapons like robot aircraft, cruise missiles, and fuel air explosives blurred lines between nuclear and conventional weapons, while improvements in weapons accuracy and the computer revolution offered ways to compensate for what some observers saw as a growing American aversion to war losses. Whether stemming from such causes as television bringing immediate color images of war to mass audiences, health care advances, urbanization, and shrinking family size, or education altering relationships between parents and children, such attitudes were visible long before the Vietnam War, as were PGMs and "unmanned aerial vehicles" (UAVs) which dated back to World War I. Several types were used in combat in World War II,[114] and although senior airmen often resisted using them, "smart bombs" appeared in newsreels from the late 1940s onward. Millions saw their effects on television in 1972 as they destroyed point targets, including bridges that had remained unscathed after hundreds of attacks with "dumb" bombs. In the early 1970s, concern for public reaction to battle losses during the Vietnam War led the Army to begin developing an "Electronic Battlefield" to minimize casualties. In 1973, in the Yom Kippur War, the effects of new types of "smart" weapons on a large scale produced stunning losses of aircraft and tanks to simple wire-guided missiles and other smart ordnance, while in 1982, in the Falklands War, the Argentine air force's Exocet missiles wreaked havoc on the Royal Navy's invasion armada.

Despite those trends, airmen took heart from the fact that "Stealth" techniques of blinding radars, PGMs, and UAVs were allowing aircraft to carry out missions more safely, with a lowered risk of downed aircrews being tortured or held hostage, a longstanding problem in aerial warfare highlighted during the Paris Peace Talks of 1969–1973. Although some innovations produced an inadvertent humanitarian effect by allowing air forces to wield power accurately and with less "collateral damage," the new weapons systems were developed to wage nuclear or conventional warfare, not to enforce peace or counter aggression in a "police" mode. Nevertheless, during the nearly two decades that elapsed before American forces used the new arsenal in a major combat operation, Operation DESERT STORM—the Gulf War of 1991, the image of international peace keeping maintained some momentum in American popular culture. In *Star Trek*, a science-fiction television series, for example, the Starship *Enterprise* cruised the galaxies under the authority of an interplanetary federation whose logo resembled the UN's. *Enterprise*'s crew faced the paradox identified by Senator Lodge in the early twentieth century in respect to an IPF—although heavily armed, it operated under a "prime directive" of "hands-off." The IPF and IAF concepts also continued to appear in defense analytics. During the detente/ *razriadka* interlude of U.S.–Soviet accommodation in the mid- and late 1970s, some theorists offered new versions of an IPF analogous to Federal marshals and the Royal Canadian Mounted Police. Like many earlier proponents, they hoped that creating such a force, however small, would lead to greater things. They were also confident that moral suasion and artful propaganda would change world opinion and allow "men of good will" to nudge the international system toward resolving conflicts by peaceful means. Notional modalities included sophisticated monitoring technology; increasingly accurate weaponry; reactive policing;[115] subtle and passive stratagems and methods such as reconnaissance, intimidation, and propaganda;[116] and the concept of "civilian-based defense."[117]

Despite such fitful interest, however, IPF and IAF schemes drifted back to the outer margins of practical politics, foreign policy, and military theoretics by the late 1970s, as the Cold War, which had seemed to be ebbing over the previous decade, began to intensify once again. America's military establishment and its will to intervene shrank in the wake of Vietnam, the Soviets bolstered their forces in central Europe and then invaded Afghanistan. The prolonged humiliation following the seizure of the U.S. embassy in Teheran produced a major shift in public opinion that brought Ronald Reagan to office. As the superpowers reverted to mutual demonization, and the United States began rearming on a major scale, a distinguished political theorist expressed a sentiment shared by the elites of many nations, that creating "a world police force capable of handling superpowers" was "almost inconceivable."[118] That view looked past Reagan's enthusiasm for such structures and the perplexing fact that both the United States and the USSR had been stymied and humbled by numerically and technically inferior foes. Again, although imperial powers throughout history often suffered similar setbacks, the recurrent stumbling of the superpowers' in limited wars raised questions about the adequacy of the calculus of military, economic, and political power. Despite soaring investments in intelligence gathering and analysis, the mapping of the future remained closer to metaphysics and soothsaying than to science. That chronic unpredictability was highlighted when the shah's regime was toppled in 1979 when the Russians withdrew from Afghanistan, the Soviet system crumbled in 1989, and when Iraq seized Kuwait in late 1990.

In only a slightly less stunning surprise, those events brightened the faded images of the IPF and IAF, and brought them seemingly closer to realization than ever before.[119] During the Cold War, it was often unclear how much a particular clash was a byproduct of Russian–American friction rather than local factors.[120] During the previous five decades, observers had often ascribed the ongoing turmoil to superpower rivalry preventing the UN from forming an effective standing force.[120] Many hoped when the Berlin Wall fell in 1989 that it presaged a major decline in limited wars. A trace of the optimism of 1943–1944 flavored George H. W. Bush's invoking of H. G. Wells' "New World Order" rubric as he assembled an armed coalition to counter Saddam Hussein's invasion of Kuwait.[121] Amid that flurry of diplomatic fence building, the UN's Military Staff Committee was briefly reactivated to gain Russia's agreement not to oppose the ad hoc alliance's use of military force against Iraq. That diplomatic maneuver, however, did not mask the fact that the Gulf War, like the Korean War, was a coalition assembled and led by the United States. The UN's authority over military operations was as flimsy as in that earlier conflict,[122] and the Gulf War obviously served American national interest.[123] But the prospect of reviving UN peace-keeping mechanisms negated by the Cold War was soon confounded,[124] and the momentary surge of hope bordering on euphoria subsided.

Although the collapse of superpower polarity did not end the pandemic of conflicts, the air elements assembled in Operation DESERT STORM came closer to an IAF in structure than any previous multinational force, while the IAF and IPF themes resonated briefly. A defense vendor's advertisement displayed a notional "World Air Force—For Peace" shoulder patch,[125] an article in *American Legion Magazine* called for a "world police force—with allegiance only to the United Nations—that does not have to await the U.S. cavalry,"[126] and a Rand study of the Gulf air war was entitled *A League of Airmen*.[127] Beyond the realm of popular culture, the use of weapons like PGMs, UAVs, and carbon filament short-circuiting bombs during the Gulf War gave new substance to the image of using airpower as a "scalpel" and serving as a nemesis of aggression in the manner of an IAF. Highly publicized and heavily censored accounts of the air war in Southwest Asia led airmen and their advocates to wax enthusiastic about using the "new technologies" produced by a "Revolution in Military Affairs" to "project" military power "surgically"—targeting weapons precisely while inflicting minimum "collateral damage," that is, civilian deaths and incidental damage.[128] Those rising expectations converged with a wave of proposals for creating a "new UN police force" that gained further momentum when UN General Secretary Boutros Boutros-Ghali proposed creating a fire brigade of "combat-ready units provided by member states" to "fill the gap" between "lightly armed UN peacekeeping units," as well as full-scale forces like those that served under the UN flag and wore UN medals during the Korean War.[129] Like many architects of such models, Boutros-Ghali proposed that member nations maintain contingents of 1,000 ready to deploy on twenty-four hours' notice under the provisions of Article 43 of the United Nations Charter. Those would be funded by an excise tax system designed to deal with some member nations' uneven contributions—including those of the United States.[130]

While some expected the end of the Cold War to turn the clock back to 1945 and allow a fresh start in building a strong UN, others opposed international structures on philosophical grounds—or saw them as sinister precursors of Orwellian world government.[131] President Reagan, in one of his last public statements, expressed hope that "the responsible powers of the world" might unite "to impose civilized standards on those who flout every measure of human decency" through a revitalized UN and a redirection of NATO. In keeping with his doctrine of supporting anti-Communist insurgencies, it echoed Reagan's youthful support of World Federalism, and his call, in 1992 at Oxford University, of the creation of "a standing UN force—an army of conscience," to provide "a humanitarian velvet glove" with "a steel fist of military force."[132] Another surge of optimism in the autumn of 1993 subsided quickly when U.S. Special Forces in Somalia under UN auspices were roughly handled by local dissidents and images of the debacle were telecast throughout the world. The growing sense of frustration was heightened by the realization that although the Gulf War had removed Iraqi forces from Kuwait, Hussein remained in power after President Bush and Prime Minister Thatcher left office. With nationalism and sovereignty blooming afresh in several settings, cynics and disappointed visionaries alike deemed the "new world order" a new world chaos.[133]

American policy regarding the structuring of international security mechanisms remained as ambiguous and schizoid as it had been since Versailles. In the mid-1990s, stung by setbacks, the Clinton administration reversed course as the Republicans regained control of Congress after a generation in the wilderness. The United States remained in arrears of its UN dues and helped to elbow Boutros Boutros-Ghali and his ambitious global collective security plans off-stage but wired the UN into American interventions in Haiti and the Balkans. As the apposite ideals of self-determination and collective security that bedeviled Woodrow

Wilson reappeared,[134] France reprised its old role in January 1992, when President Francois Mitterand offered to place 1,000 French troops under the authority of the UN's Military Staff Committee.

The "public security gap" that appeared following international intervention in Bosnia led to fresh designs for maintaining law and order that looked much like older models, such as proposals for forming United Nations "civpol"—civilian police—contingents to maintain public order during the "transitional phase" as security functions were shifted from military to civilian authorities.[135] Falling back on regional arrangements, the major Western powers formed an "International Police Task Force" (IPTF) under the banner of a new, expansive, and ultimately aggressive NATO. While not an IPF in the classic sense, those police elements imposed some momentary stability in that lurid *Burgerkrieg*.[136] In a growing thicket of acronyms in the early 1990s, the UN's Department of Peacekeeping Operations (DPKO) undertook several small initiatives in international police training, oversight, and coordination, rekindling the hopes of those who advocated strengthening the UN's military capacity and authority.[137] On a parallel track, in 1995 the Organization for Security and Coordination in Europe (OSCE) was formed from the two-decade-old Conference on Security and Cooperation in Europe, followed by such structures as United Nations Protection Force (UNPROFOR) and Implementation Force (IFOR) created when the global pot boiled over in several regions later in the decade.[138] The overall trend in UN involvement in such activities, however, was downward, with a slightly more than 80 percent decline from 1993 to 1999.[139]

As the twentieth century ended, Saddam Hussein's attempts to thwart UN–sanctioned weapons inspection led to an ongoing deadly minuet as his violations of the 1991 ceasefire agreement were countered by occasional air strikes by American and British aircraft in southern Iraq. At the same time, the failure of multinational ground forces in a weak IPF configuration under the UN's aegis to stabilize the situation in Bosnia led NATO to create a kind of IAF to coerce Serbia. Beginning in June 1998, the thirteen members' air forces mounted a massive show of force to intimidate the Serbs in Kosovo when Slobodan Milosevic rejected the Rambouillet agreement. That coercive but passive demonstration took place as a debate mounted in U.S. defense circles over optimal methods of employing military force. A primary pivot of contention was the "Halt Phase Strategy/Doctrine"—"Halt" for short—based on U.S. Air Force claims that it could most rapidly and effectively deal with an aggressor pending the arrival of Army and Marine Corps forces.

A linked controversy had been under way since the Gulf War in defense circles over how much the destruction of Iraqi forces in DESERT STORM was due to Allied Coalition—mainly American—air operations versus the ground offensive. Those urging that the first phase of future

American armed intervention should be assigned to the Air Force cited as positive examples the Battle of Khafji in 1991, and Operation DELIBERATE FORCE, which preceded the Serbs' agreement to enter the Bosnian negotiations in 1995. Critics cited the difficulty of sorting out air and ground forces' respective contributions in both cases, due to their very close interaction in the dynamics of battle.[140] That debate was engulfed by the flow of events as air power became NATO's primary modality in the Balkans, first, as noted earlier, as a means of intimidation in 1998, and then, in the spring of 1999, in subjecting Serbia to an air offensive that grew in stages to exceed DESERT STORM in length and intensity, including a long series of major attacks by a mix of UAVs, cruise missiles, and conventional and Stealth aircraft, from both distant and nearby bases.[141] Once again, things seemed clearer during the operation than afterward, and it led to further debates in defense circles over whether military force should be used with or without ground troops and when potential hostages were at risk.[142]

Amid all the tumult, there was no direct reference to "air policing," even though UN officials, diplomats, military commanders, and journalists were grappling with the complex problems of providing police services in Kosovo, where tens of thousands of NATO troops and some Russian ground units were deployed as police, as well as a 3,000-strong "international police force." At the same time, away from the main arena, the European Union created its own military force with a general staff and headquarters,[143] a move that some Americans approved[144] despite continuing uncertainty as to whether the United States, under such new rubrics as "coercive inducement," was becoming what William Appleman Williams called the "imperial police power," a role that it had apparently abandoned at the end of the Vietnam War."[145] As the UN police contingent was swamped, the main burden of maintaining local security fell on the 50,000 NATO troops in the region. The wrangling of NATO and UN authorities over imposing order touched on some classic IPF dilemmas, like defining crime, control of the police, and keeping the training and equipping of law enforcement elements from enhancing repressive regimes' ability to coerce and punish under rubrics like civil order and security.[146]

The intricacies of unilateral and multinational intervention, including the lack of a firm definition of "peacekeeping," a word that does not appear in the UN Charter, had also been highlighted in the long series of small civil wars in Beirut in the early 1980s, and in Haiti and Somalia a decade later.[147] It was not clear, as it had not been between the world wars, whether the term "police" in "international police force" meant police in the civil sense, or the use of military power as Teddy Roosevelt and Henry Wallace meant it to be. On the American domestic scene, the divergence of roles and expectations between constabulary and sustained combat functions, as well as extended foreign deployments, created increasing cross-strains in the U.S. armed forces, active and reserve.[148] The Somalian debacle led President Clinton to issue a directive in May 1994 that set guidelines for applying force in peace-keeping operations, including criteria for determining if U.S. forces should be committed under UN authority,[149] matters that were also addressed in the International Police Task Force ground rules laid down during the Balkan peace talks at Dayton, Ohio, in late 1995.[150] In the Balkan intervention, in contrast with earlier cases, neither tight rules of engagement, nor the shunting American of troops from UN to NATO control had much apparent specific affect on American attitudes. The traditional ambivalence of elites appeared in public opinion polls showing overall support for UN–based intervention, but only one of three Americans favoring the dispatch of U.S. forces to Bosnia,[151] and revealing a split similar to that of World War II between public and congressional views on internationalism.[152]

In the Balkans in the late 1990s, as in the 1940s, collective will had little bearing on the flow of events. The decline in American official support for the United Nations for a generation, brought the United States further in arrears to the world organization than any other country[153] and below the top ten among the 71 of 185 UN member nations whose forces were deployed in peace-keeping roles.[154]

Nor had the age-old question of sovereignty been resolved. Despite continuing attacks on the UN from various quarters[155] and the trend of smaller nations picking up the slack in peace-keeping missions in various parts of the world,[156] an American analyst saw sovereignty as still hale and hearty from a legal standpoint and the UN's lack of "political will and the resources to deal with these matters on a consistent basis throughout the globe" as a favorable trend.[157] In the mid-1990s, rising fears on the American far Right about the apparent expansion in UN authority and threats to U.S. sovereignty were highlighted by the court-martial of a soldier for refusing to serve under UN auspices.[158] On the other side of the political aisle, the American Bar Association supported Brian Urquhart's call for "the creation of a standby military force of units from the national military forces of [UN] member states, pursuant to Article 43 of the United Nations Charter."[159]

Polarity on those issues was less well defined than it had been during World War II. Moderates as well as "militia–patriot" groups and neoisolationists[160] opposed U.S. forces serving under foreign commanders,[161] political columnist Patrick Buchanan carried a neoisolationist banner as a presidential candidate,[162] and Henry Kissinger judged Woodrow Wilson as having gone "totally wrong . . . in the idea of collective security."[163] Such anxiety over encroaching foreign dominion or the schemes of elites was not new to the American scene—nor exclusively American. The North American Free Trade Agreement (NAFTA) did not blunt the nationalistic ardor of any of the signatory nations. Mexican Army leaders, for example, reportedly rejected American offers to cooperate on defense matters, out of concern that "military integration and the use of Mexico's armed forces" as part of an international "super police" under U.S. command would lead to the "overriding [of] Mexican sovereignty."[164]

The perennial schizoid split between American policy and public attitudes was not only visible in the Clinton administration's backing away from the United Nations,[165] but also in its cutting foreign aid over 50 percent in the 1990s at the same time that some dormant and seemingly extinct internationalist mechanisms were being revived. In the late 1990s, for example, U.S. and Russian aircraft began reciprocal "Open Skies" high-altitude reconnaissance overflights, a concept originally proposed by the Eisenhower administration to detect what Kenneth Boulding defined as "young conflict processes."[166] Such concerns remained well out of public view throughout the 1990s in the United States. As the nation once again turned inward after major wars, network television news coverage "dumbed down," newspaper readership declined,[167] and foreign policy was overshadowed by such domestic concerns as the soaring stock market, presidential scandals, the year 2000 (Y2K), crime, health care, education, prisons, drugs, et al. Although there was little evidence of the IPF and IAF concepts in public discourse, such hoary rubrics as "New World Order" appeared from time to time as did Franklin Roosevelt's vision of a "sheriff, forging posses to deal with problems as they emerge" in the form of ad hoc "coalitions of the willing" comprised of national contingents.[168] The Clark-Sohn plan still had its admirers,[169] the RAF's "Imperial Policing" model for "constabulary operations" was dusted off and rolled out occasionally,[170] and some observers claimed to see clear patterns amid the post–Cold War turbulence. The *Gulf War Air Power Study*'s editor, for example, described air operations in the Balkans as "the kind of imperial policing that the American military is increasingly going to be called upon to do" in "the post-cold-war world,"[171] and N. D. White asserted on the eve of the Serbian air campaign that "inconsistent responses of the world community . . . illustrate that as yet the United Nations cannot claim to embody the United Nations ideal."[172]

As the new century and millennium began, some observers saw upper-tier powers' involvement in the Balkans, Iraq, Ulster, Timor, and Chechnya as regressive. True or not, those and other flickering conflicts showed that brushfire wars had survived the Cold War. Even though concepts and theories designed to deal with them still abounded, the half-century of experience produced no reliable techniques for anticipating and dealing with them quickly and cheaply.[173] Intervention policy often looked like old wine in old bottles. As the bombing of Serbia began, for example, Clinton administration officials claimed that leaving such unfolding horrors unchecked risked repeating the Western democracies' failure to block Hitler's rise, echoing similar assertions by the Eden government during the Suez affair, the Johnson administration's rationale for intervening in Vietnam, and the Bush administration's rhetoric in confronting Saddam Hussein. Other old refrains from the interval between the world wars included uncertainty about air bombing's efficacy, claims of moral superiority, and sharp differences over military leaders' competency[174] and American involvement and disengagement.[175] Beyond assertions that using military power to impose peace and counter aggression was a form of neoimperialism[176] were other rough parallels to the past in the Serbian air campaign including uncertainties over air power's ability to produce quick, cheap, clear results;[177] the targeted regime held on to power, and the allies wielding the air weapon disagreed regarding optimum methods.[178]

Beyond debates over whether a multinational bombing campaign had been necessary to halt aggression in Kosovo,[179] the IPF and IAF rubrics had all but disappeared from the worlds of theory,[180] policy, and practice, aside from an occasional mention by analysts and military professionals.[181] That seemed somewhat curious, considering *de facto* traces of those concepts in the Balkan air campaigns throughout the 1990s and the involvement of several nations' air forces. As in Korea, the Gulf War, intentionally or not, was fought under collective security with the United States providing the lion's—or eagle's—share of forces. That could be ascribed to American wealth and power, or European members' relatively low defense expenditures,[182] or less obvious factors like the uneven or meshing of systems among NATO forces, despite half a century of "interoperability," which produced problems with command and control, basing access, and security.[183] At the end of the bombing of Serbia, as in previous air campaigns, there was substantial disagreement regarding results. American airmen claimed a Mitchellian victory, while critics of the Serbian bombing argued that even though no airmen were lost, and the Milosevic regime signed an agreement and backed off, hundreds—or thousands—of civilians had been killed, while the air attack failed to halt "ethnic cleansing" in Kosovo and perhaps triggered or intensified it.[184] And in the way of an ironic aside, at the same time that some advocates of bombing Serbia cited the failure to nip Hitler in the bud as justification for applying coercive military force, historians were debating whether military power—especially air attacks—should have been used against the Holocaust infrastructure.[185]

Looking past those recent events, it is perplexing to note how many of the basic problems that confronted Henry IV half a millennium ago, when he attempted to design an internal peace enforcement system, were visible when the twenty-first century dawned, including contradictions between practical politics and spiritual and idealistic visions of a peaceful world order, the dilemmas of using force against force and of great nations coercing smaller states, and the unresolved if not irresolvable tensions between the gleaming grail of the ideal of international order and the unyielding armor plate of sovereignty. Architects of peace systems still draw strained analogies between local security and military coercion, and attempts to enforce peace remain indistinguishable from ploys to maintain the status quo. Most important, the potential costs of leaving those quandaries unresolved have been rising steeply for the last 150 years. It may be that the IPF and IAF concepts have never been more than keys on a faery ring of fanciful nostrums, like arbitration and arms control, none of which could have been fitted into the lock of peace under any circumstances. As abstract and academic as that may seem, the quandary that it raises is deadly serious. Since no scheme for keeping the peace on a major scale has worked so far, is that because we failed to find the right one, or because it is destined to always lie out of reach?

How is it possible to be certain? If it were proved to be the latter, then things would be no worse than they have been all along in the domain of statecraft, a notional museum of which would include such grisly exhibit exhibits as the Tai p'ing Rebellion, the world wars of the modern era, Cambodia, and Rwanda. The question of "maintaining" peace would remain enshrouded as it has been over the centuries, as a fantasy like Arcadia, Atlantis, Camelot, and Elysium, with the IAF relegated to a piece of apparatus used in wishful thinking exercises. But if the dilemma was found to be tractable, then the elements of mechanism, methods, and epistemology would become very concrete and practical, along with the IAF concept's viability. As we have seen, it is alive, ipso facto, in faint and scattered fragments, in certain situations, but far from being the solid edifice that so many have tried to design and so few have tried to actually build. There is no formal ongoing structure nor doctrine with footings in international law in sight, nor a prospect of one being erected. And very few are aware of even the rudiments of that intricate architectural history, and, therefore, of precedents.

Despite its moribund state, considering the IAF concept as a gedanken experiment does heighten awareness of how, as Mario Puzo suggested obliquely in *The Godfather*, the international "system" lies closer in essence to being an ecology of gangsterism than the elegant facades and semantics of statecraft and academic analysis suggest. Some "key players" are highly intelligent, genteel, courteous, and personally generous and kind; others are pathologically cruel and selfish; and many blends of those polarities lie on the spectrum between. Beyond their style and motives, all are oriented to violence and compelled by the implicit rules of the "game" of international relations to be alert to and ready to use force, albeit to widely varying degrees. Franklin Roosevelt tacitly admitted that when he referred to "the Four Policeman" and the posse model, the latter was closer to the mark inasmuch as it describes an ad hoc response in a lawless setting. Such active aggregations were formed—and virtually always encompassed traditional foes—to various purposes again and again, to constrain or confront such monarchs and dictators as Louis XIV, Napoleon, the Central Powers, the Axis, Kim Il Sung, and Saddam Hussein. But there are no clear patterns of appropriateness and proportionality, nor a gradient or scale by which we can appraise those instances, beyond drawing broad inferences. Nor is it clear what the costs have been of leaving these matters to questionable processes and turns of chance. But it is plain enough that, so far, maintaining what is so often and so loosely labeled "international order" has been left to chance.

As gloomy a prognosis as this may suggest, there is some hope, however dim. Social scientists are on the track of such problems, even if military and diplomatic history still lie closer to art than science. The cyber age may very well provide better tools and a clearer view of the potential risks and costs of failing to confront this matter head on, and more important, furnish ways of muzzling the Cerberus of sovereignty, nationalism, and war. It may even become possible to "diagnose" and "treat" well in advance of major pathology. Perhaps it will show that the analogy of civilian police that underlies the IAF and IPF concepts is not appropriate and that such metaphors as fire prevention or emergency medical response are more apt images of maintaining peace through direct, multinational action. What is clear is that, like crime and disease, there is no prospect of the problem going away, nor will the hazards of persisting in our heedlessness and leaving our fate to the fall of metaphorical dice in the form of ad hoc responses.

NOTES

1. Quoted Inis L. Claude Jr., "The United Nations and the Use of Force," *International Conciliation* 532:2 (March 1961): 330.

2. United Nations General Secretariat, "Agenda for Peace," quoted in *Canadian Centre for Global Security Communique*, November 1992, p. 3.

3. For a list of such undertakings from 1948 to the mid-1990s, see Myron H. Nordquist, *What Color Helmet? Reforming Security Council Peacekeeping Mandates* (Newport, R. I.: Naval War College, 1997) [Newport Paper No. 12], pp. 19–33.

4. Insurgencies included the Hukbahalap rebellion in the Philippines in the late 1940s and early 1950s; the Malayan "emergency" of 1948–1962; the Indonesian rebellion 1945–1947; the Algerian War, 1954–1962; the Mau-Mau uprising in Kenya 1950–1954; Cyprus, 1952–1964; and the Suez Crisis, 1956. Wars included those between India and Pakistan in 1948, 1956, 1964, and 1981; the Arab-Israeli Wars of 1948, 1956, 1967, and 1973, and the Lebanon campaign of the early 1980s; the Iran-Iraq War, 1980–1988; the Korean War 1950–1953; the 2nd Indo-China War, 1956–1975; the Falklands War of 1982; the Russian incursion into Afghan-

istan 1979–1989; the wars of Soviet devolution, from 1989 to the present and the Balkan wars of the 1990s. Beyond those were hundreds of conflicts at the lower end of the spectrum of intensity.

5. Brian Urquhart, "Looking for the Sheriff," *The New York Review of Books* 45:12 (July 16, 1998): 48.

6. Frank G. Williston and Linden A. Mander, "The Problem of Security," in Joseph B. Harrison, Linden A. Mander, and Nathaneal H. Engle, eds., *If Men Want Peace: The Mandates of World Order* (New York: Macmillan, 1946), pp. 29–30.

7. Henry A. Wallace, *The American Choice* (New York: Reynal and Hitchcock, 1948), p. 33.

8. Public Law 264, 79th Congress, 1st session, December 1946.

9. George Fielding Eliot, *The Strength We Need: A Military Program for America Pending Peace* (New York: Viking, 1946), esp. pp. 66–93; for another perspective on the naval police role, see George Stitt, "The Functions of the Navy in Support of the Security Council," *The Fighting Forces* 23:4 (October 1946): 190–191.

10. Ibid., pp. 588–589, 591.

11. Ibid., pp. 592–593.

12. An account of that metamorphosis is Cord Meyer Jr., *Facing Reality: From World Federalism to the CIA* (New York: Harper and Row, 1980), pp. 44–56.

13. Cord Meyer Jr., *Peace or Anarchy* (Boston: Little Brown, 1947), pp. 143–183.

14. Emery Reves, *The Anatomy of Peace* (New York: Harper and Brothers, 1946), p. 29; also see pp. 184–199.

15. Perhaps the most exhaustive and technically sophisticated collective effort at that time was the August 1946 issue of *Yale Law Review* which included essays by Edwin Borchard, Leland Goodrich, Robert Johansen, Hans Kelsen, Harold Lasswell, Stefan Possony, and Quincy Wright.

16. Sigrid Arne, *United Nations Primer* (New York: Rinehart and Co., 1948), p. 265.

17. Meyer, *Peace or Anarchy*, p. 207.

18. For the progress of those efforts, see the Brookings Institution's *Major Problems of U.S. Foreign Policy: A Study Guide* series for 1948 through 1951; also see Andrew Martin, *Collective Security: A Progress Report* (New York: UNESCO, 1951), pp. 166–167.

19. James F. Byrnes, *Speaking Frankly* (New York: Harper and Brothers, 1947), pp. 303, 314.

20. National Air Council, *Peace Through Air Power* (Washington, D.C.: Air Power League, 1946), p. 31.

21. Carl A. Spaatz, "The Truth About Your Air Force," *Coronet* 22:5 (September 1947): 10–16.

22. Carl Spaatz, "Strategic Air Power: Fulfillment of a Concept," *Foreign Affairs* 24:2 (April 1946): 396.

23. Bernard Brodie, *Strategy in the Missile Age* (Princeton: Princeton University Press, 1959), p. 293.

24. John Erickson, "The Soviet View of Deterrence: A General Survey," *Survival* 24:6 (November/December 1982): 242–249.

25. R. Fulljames, "The Problem of Security from the Fear of War in the Atomic Age," *Fighting Forces* 24:2 (June 1947): 86.

26. J. M. Spaight, *Air Power Can Disarm* (London: Pitman and Sons, 1948), pp. 163, 168.

27. Grenville Clark and Lewis B. Sohn, *World Peace Through World Law* (Cambridge: Harvard University Press, 1958), esp. pp. 300–303.

28. E.g., see Lincoln Bloomfield, ed., *International Military Forces* (Boston: Little Brown, 1964); and idem, "International Force: A Symposium," *International Organization* 17:2 (September 1963).

29. E.g., see *A Constitution for the World* (Santa Barbara: Center for Democratic Institutions, 1965), esp. p. 49.

30. Daniel S. Cheever and H. Field Haviland, *Organizing for Peace: International Organization in World Affairs* (Boston: Houghton Mifflin, 1954), p. 484; a succinct overview is Clyde Eagleton, *International Government* (New York: Ronald Press, 1948, pp. 446–454.

31. Elliott R. Goodman, *The Soviet Design for a World State* (New York: Columbia University Press, 1960), p. 425.

32. Quincy Wright, "Aviation and World Politics," *Air Affairs* 1:1 (n.d) [1946]: 104.

33. E.g., see Sir Gerald Charles Dickens, *Bombing Strategy: The Fallacy of Total War* (London: Low, 1947).

34. David MacIsaac, ed., *The United States Strategic Bombing Survey*, 10 Vols. (New York: Garland Publishing Co., 1975).

35. Stefan T. Possony, *Strategic Air Power· The Pattern of Dynamic Security* (Washington, D.C.: Infantry Journal Press, 1949), esp. pp. 243–291.

36. Dale O. Smith, "Air Power as Peace Power," *Air University Quarterly Review* 3:3 (Summer 1949): 13, 17.

37. That law also created the National Military Establishment, the Office of the Secretary of Defense, a new statutory definition of the Joint Chiefs of Staff, the National Security Council, the Director of Central Intelligence, and the Central Intelligence Agency.

38. Vernon Nash, *The World Must be Governed* (New York: Harper and Brothers, 1949), pp. 142, 193–195; also see Edith Wynner, *World Federal Government: Why? What? How?* (Afton, N.Y.: Fedonit Press, 1954).

39. Carl Spaatz, "Air Power in the Atomic Age," *Collier's* 116 (December 8, 1945): 11ff.

40. E.g., see Linden A. Mander, *Foundations of Modern World Society* (Stanford: Stanford University Press, 1948), pp. 88, 659–797.

41. Martin, *Collective Security*, p. 17.

42. E.g., see Eugene M. Emme, *National Air Power and International Politics: A Select Bibliography* (Maxwell Air Force Base: Historical Division, Department of the Air Force Library, [January] 1950).

43. Quoted in Frederick L. Schuman, *The Commonwealth of Man: An Inquiry into Power Politics and World Government* (New York: Alfred A. Knopf, 1952), p. 411.

44. Denna Fleming, *The United States and World Organization, 1920–1993* (New York: Columbia University Press, 1938), pp. 536–537.

45. Werner Levi, *Fundamentals of World Organization* (Minneapolis: University of Minnesota Press, 1950), p. 10.

46. Stanley Meisler, *United Nations: The First Fifty Years* (New York: Atlantic Monthly Press, 1995), p. 63.

47. Estimates range from 300,000 to two million.

48. Tom Slick, *The Last Great Hope* (San Antonio: Naylor Co., 1951), p. 39.

49. E.g., Gerald J. Mangone, *The Idea and Practice of World Government* (New York: Columbia University Press, 1951), pp. 55–64.

50. Some United Nations' information material aimed at broad public readership has not mentioned the military dimension. E.g., see *How the United Nations Began* (New York: United Nations, 1951) [UN Publication 1951–1-1].

51. *United States Air Force Operations in the Korean Conflict, 25 June 1950–1 November 1950* (Maxwell Air Force Base: Department of the Air Force, 1952), p. 84.

52. Malvern Lumsdern, *Incendiary Weapons* (Cambridge: MIT Press, 1974), p. 45.

53. Charles Seymour, *Geography, Justice and Politics at the Paris Peace Conference* (New York: American Geographic Society, 1951), p. 24.

54. Inis Claude Jr., "The Police Function of the United Nations," in Robert Theobald, ed., *The UN and Its Future* (New York: H. H. Wilson, 1963), p. 96.

55. Schuman, *Commonwealth of Man*, pp. 407, 414, 492. For details of the British scheme, see PRO AIR 75/72, "A European Air Force: Notes by the Chief of Air Staff," January 2, 1952.

56. Thomas K. Finletter, *Power and Policy: U.S. Foreign Policy and Military Power in the Hydrogen Age* (New York: Harcourt and Brace, 1954), p. 319.

57. E.g., the Anglo-American-Canadian "Atomic Declaration" and Big Four Foreign Ministers' agreement of 1945; the Acheson-Lilienthal report; the Baruch Plan; Truman's nuclear disarmament and British-French-American proposals to the General Assembly for a "general political settlement" to Cold War differences in November 1951; the Rapacki Plan; the Vishinsky Proposal; the Bolte Plan; the Zorin-McCloy statement, etc.

58. Vannevar Bush, "We're Arming for the Wrong War," *Nation's Business*, August 1955, p. 21.

59. A detailed account of the Ely mission is Jules Roy, *The Battle of Dienbienphu*, trans. Robert Baldick (New York: Harper and Row, 1965).

60. E.g., see Willard N. Hogan, *International Conflict and Collective Security: The Principle of Concern in International Organization* (Lexington: University of Kentucky Press, 1955).

61. Asher Lee, *Air Power* (New York: Praeger, 1955), p. 45.

62. W. D. Puleston, *The Influence of Force in Foreign Relations* (Toronto: Van Nostrand, 1955), pp. 197–214.

63. E.g., the Navy's SEAL teams and the US Air Force's Air Commandos were formed in 1963, while the US Army's Special Forces, created in 1953, were tripled in strength.

64. John W. Burton, *Peace Theory: Preconditions of Disarmament* (New York: Knopf, 1962), pp. 83–92, 99, 166, 172, 197–198.

65. Inis L. Claude Jr., *Swords Into Plowshares: The Problems and Progress of International Organization* (New York: Random House, 1956), esp. pp. 250–294; and

W. H. Carter, "Thoughts on International Policy," *Contemporary Review* 94:1116 (December 1958): 313–315.

66. E.g., see Lewis B. Sohn, "The Authority of the United Nations to Establish and Maintain a Permanent United Nations Force," in Saul H. Mendlovitz, *Legal and Political Problems of World Order* (New York: The Fund for Education Concerning World Peace Through World Law, 1962), pp. 323–334.

67. For a detailed analysis, see Julius Stone, *Aggression and World Order: A Critique of United Nations Theories of Aggressions* (Berkeley: University of California Press, 1958), pp. 176–200.

68. Bruce Pirnie and William E. Simons, *Soldiers for Peace: Critical Operational Issues* (Santa Monica, Calif.: Rand Corporation, 1996), p. 71.

69. Richard N. Gardner, *In Pursuit of World Order: U.S. Foreign Policy and International Organizations* (New York: Praeger, 1964), p. 80.

70. Quoted in Bill Kauffman, *America First! Its History, Culture, and Politics* (Amherst, N.Y.: Prometheus Books, 1995), p. 63.

71. E.g., see excerpts from "The Truth About Bob Jones University," in "The Tower of Babel: Bob Jones's Dating Tips," *New York Times*, March 5, 2000, sec. 4, p. 5, and "University Drops Ban on Interracial Dating," *Houston Chronicle*, March 4, 2000, p. 21A [Associated Press release].

72. Henry Cabot Lodge Jr., "Fruitful Approach to a Prolonged Peace," *Vital Speeches of the Day* 9:22 (September 1, 1943): 682–688.

73. Amitai Etzioni, *Winning Without War* (Garden City, N.Y.: Anchor Books, 1965), p. 201.

74. Herbert C. Kelman, "Institutionalizing Military Force," in Quincy Wright, William H. Evans, and Morton Deutsch, eds., *Preventing World War II: Some Proposals* (New York: Simon and Schuster, 1962), pp. 109, 118–119.

75. F. H. Hinsley, *Power and the Pursuit of Peace: Theory and Practice in the History of Relations Between States* (Cambridge: Cambridge University Press, 1963), p. 345.

76. Arthur I. Waskow, "Quis Custodiet? Controlling the Police in a Disarmed World," and Hans Morgenthau, "Political Conditions for a Force," in Richard A. Falk and Saul Mendlovitz, eds., *Strategy of World Order* (New York: World Law Fund, 1966), 3:620–658 and 660–667 respectively; James Stegenga, "United Nations Peacekeeping: Patterns and Prospects," in Robert S. Wood, ed., *The Process of International Organization* (New York: Random House, 1971), p. 300; and Victor H. Wallace, *Paths to Peace: A Study of War: Its Cause and Prevention* (London: Cambridge University Press, 1957), p. 366.

77. James M. Straubel, "Peace on Earth—Controlled from Space," *Air Force*, March 1960, reprinted in John F. Loosbrock and Richard M. Skinner, *The Wild Blue: The Story of American Air Power* (New York: G. P. Putnam's, 1961), p. 618.

78. Claude, "Police Function," p. 96.

79. Kai Bird and John J. McCloy, *The Chairman: The Making of the American Establishment* (New York: Simon and Schuster, 1992), p. 515; for a text of the agreement, see Falk and Mendlovitz, *Strategy of World Order*, 3:280–282.

80. Claude, "United Nations and the Use of Force," pp. 247, 338.

81. Philip E. Jacob, Alexine E. Atherton, and Arthur M. Wallenstein, *The Dynamics of International Organization* (Homewood, Ill.: The Dorsey Press, 1972), pp. 81–109.

82. David Mitray, *A Working Peace System* (Chicago: Quadrangle Books, 1966), p. 61.

83. E.g., see Leland M. Goodrich and David A. Kay, eds., *International Organization: Politics and Processes* (Madison: University of Wisconsin Press, 1973).

84. Burton, *Peace Theory*, pp. 86–92; and Hans Morgenthau, "The Political Conditions for an International Police Force," *International Organization* 17:2 (September 1963): 400–402.

85. E.g., see Committee on Study of Legal Problems of the United Nations, "Should the Laws of War Apply to United Nations Enforcement Action?" in Falk and Mendlovitz, *Strategy of World Order*, 3:70–75, as well as many articles on the subject in vols. 3 and 4.

86. Walter Millis and James Reed, *The Abolition of War* (New York: Macmillan, 1963), p. 213; Walter Millis, *An End to Arms* (New York: Atheneum, 1965), pp. 192, 193, 202, 208, 258–260; also see his "The Role of Police Forces in Response to Violations," as well as Roger Fisher, "International Police: A Sequential Approach to Effectiveness and Control," and Hans Morgenthau, "The Impartiality of the International Police," all in Richard J. Barnett and Richard A. Falk, eds., *Security in Disarmament* (Princeton: Princeton University Press, 1965), 3:286–319, 240–285, and 320–240, respectively.

87. D. W. Bowett, "Structure and Control of United Nations Forces," in Falk and Mendlovitz, *Strategy of World Order*, 3:610.

88. For detailed accounts of ROLLING THUNDER, see James Clay Thompson, *Rolling Thunder: Understanding Policy and Program Failure* (Chapel Hill: University of North Carolina Press, 1980), and *The Pentagon Papers: The Defense Department History of United States Decision-Making on Vietnam* (Boston: Beacon Press, n.d. [1972]) Gravel Edition, 3:352–353.

89. E.g., hospitals, airfields with foreign aircraft parked on them, port facilities and ships, and antiaircraft sites on dikes.

90. E.g., see Doris Kearns, *Lyndon Johnson and the American Dream* (New York: Harper and Row, 1976), p. 263.

91. Jack Valenti, *A Very Human President* (New York: W. W. Norton, 1975), p. 224.

92. Recent perspectives are "Afterword," David M. Barrett, *Uncertain Warriors: Lyndon Johnson and His Vietnam Advisors* (Lawrence: University Press of Kansas, 1993), pp. 190–194; and Robert Buzzano, *Masters of War: Military Dissent and Politics in the Vietnam Era* (Cambridge: Cambridge University Press, 1996), pp. 298–301.

93. For a recent analysis of the question of feasibility, see Colonel Robert Schandler, "U.S. Military Victory in Vietnam: A Dangerous Illusion," in Robert S. McNamara, James G. Blight, and Robert K. Brigham, *Argument Without End: In Search of Answers to the Vietnam Tragedy* (New York: Public Affairs, 1999), p. 343.

94. See Harry Middleton, *The Compact History of the Korean War* (New York: Hawthorn Books, 1965), pp. 208–211.

95. Ralph Littauer and Norman Uphoff, eds., *The Air War in Indochina* (Boston: Beacon Press, 1972).

96. Richard Schickel, *The Disney Version* (New York: Simon and Schuster, 1968), p. 274.

97. *Impact of Air Attack in World War II—Selected Data for Civil Defense Planning Division III: Social Organization, Behavior and Morale Under Stress of Bombing* (Stanford: Stanford University Press, 1953), pp. 78–90, 116–118, 172.

98. Michael Sherry, *The Rise of American Air Power: The Creation of Armageddon* (New Haven: Yale University Press, 1987), esp. pp. 60–69.

99. For details, see *Unclassified Statistics on Southeast Asia* (Washington, D.C.: Comptroller, Office of the Secretary of Defense, 1971).

100. For a presentation of that case, see U.S. Grant Sharp, *Strategy for Defeat* (San Rafael, Calif.: Presidio Press, 1978), p. 118.

101. E.g, see Wilfred Burchett, *Vietnam North* (New York: International Publishers, 1966), pp. 170–171.

102. E.g., see Alain Jaubert, "Zapping the Viet Cong by Computer,"*New Scientist*, March 30, 1972, p. 685–688; and Arthur H. Westing and E. W. Pfeiffer, "The Cratering of Indochina," *Scientific American* 226:5 (May 1972), pp. 21–29.

103. Quincy Wright, *A Study of War* (Chicago: University of Chicago, 1969), pp. 275–278, 416–417.

104. Melvin Gurtov and Konrad Kellen, *Vietnam: Lessons and Mislessons* (Santa Monica, Calif.: Rand Corporation, 1969), pp. 14–15.

105. Pentagon Papers, 3: 282.

106. Konrad Kell, "Conversations With Enemy Soldiers in Late 1968/Early 1969: A Study of Motivation and Morale," RAND Memorandum 6131-ISA/ARPA, pp. 52–58.

107. Robert McNamara et al., *McNamara, Blight, and Brigham, Argument Without End* (New York: Public Affairs LLC, 1999), p. 4.

108. For details, see *Airpower and the Spring 1972 Invasion: U.S. Air Force Southeast Asia Monograph*, vol. 11 (Washington, D.C.: U.S. Government Printing Office, 1979); and *Unauthorized Bombing of Military Targets in North Vietnam, Hearings, Armed Services Investigatory Committee of the Committee on Armed Services, House of Representatives, 92nd Congress, 2nd Session* (Washington, D.C.: U.S. Government Printing Office, 1972).

109. For a contemporaneous perspective, see Charles K. Hopkins, "Linebacker II: A First Hand View," *Aerospace Historian* 23:3 (September 1976): 128–143.

110. Grenville Clark, *Introduction to World Peace Through World Law* (Chicago: World Without War, 1973), pp. 28–30.

111. E.g., Spain in Latin America in the 1820s, Cuba and the Philippines in 1898, and Morocco in the 1920s; Italy in Ethiopia in 1896; Britain in the American Revolution, the siege of Bhurtpore, the Sudan in the 1880s, the Boer War, the Irish rebellion of 1916, and the Suez affair; and France in Indo-China and Algeria.

112. James E. Dougherty, *Security Through World Law and World Covenant: Myth or Reality?* (Lexington, Mass.: Lexington Books, 1974), pp. 10–14.

113. Dennis J. Duncanson, Richard A. Yudkin, and Barry Zorthian, *Lessons of Vietnam: Three Interpretative Essays* (South Orange, N.J.: Institute of Far Eastern Studies, Seton Hall University, 1971), p. 31.

114. A somewhat dated overview is Roger A. Beaumont, "Rapiers and Clubs: The Strange Fitful History of Smart Bombs," *Journal of the Royal United Services Institute* 126:3 (September–October 1981): 45–50.

115. James D. Farman, *That Mad Game: War and the Chance for Peace* (New York: Charles Scribner's Sons, 1980), pp. 189–190.

116. Earl W. Foell and Richard A. Nenneman, *How Peace Came to the World* (London: MIT Press, 1986), pp. 106–107.

117. See Harry B. Hollins, Averill L. Powers, and Mark Somers, *The Conquest of War: Alternative Strategies for Global Security* (Boulder: Westview, 1989), esp. pp. 183–186.

118. E.g., Johann Galtung, *The True Worlds: A Transitional Perspective* (New York: The Free Press, 1980), p. 363; the author makes no reference to an IPF or IAF in his list of "Strategies of Demilitarization," p. 215.

119. An academic perspective on peace keeping on the eve of the Gulf War is Leland M. Goodrich and Anne P. Simmons, *The United Nations and the Maintenance of International Peace and Security* (Dordrecht, Netherlands: Martinus Nijhoff, 1987).

120. E.g., see Lincoln Bloomfield, "Arms Control and International Order," in Wood, ed., *The Process of International Organization* (New York: Random House, 1971), pp. 317–318.

121. The creep of presidential authority to control "the most important steps to war" is noted in James Grafton Rogers, *World Policing and the Court: An Inquiry Into the Powers of the President and Congress—Nine Wars and a Hundred Military Operations, 1789–1945* (Boston: World Peace Foundation, 1945), p. 86.

122. Anthony Lewis, "The New World Order," *New York Times*, May 17, 1992, p. 17E.

123. N. D. White, *Keeping the Peace: The United Nations and the Maintenance of International Peace and Security*, 2nd ed. (Manchester: Manchester University Press, 1997), p. 124.

124. Eric Schmitt, "Revived Military Panel is Seen as Largely Symbolic," *New York Times*, November 4, 1990, p. 8.

125. Advertisement by Kong Industries on the inside cover of *Aerospacio Revista Nacional*. A query of the firm by the author revealed no clear link to an actual plan or organization.

126. Timothy Stanley, *The American Legion*, March 1993, pp. 32–34, 54.

127. James A. Winfield, *A League of Airmen: U.S. Air Power in the Gulf War* (Santa Monica, Calif.: Rand Corporation, 1994).

128. E.g., Tom Clancy, "B–2 A Nightmare Come true for Tin-pot Despots," *Houston Chronicle*, July 2, 1991.

129. For a contemporaneous perspective, see Brian Urquhart, "Who Can Stop Civil Wars?" *New York Times*, December 29, 1991, p. 9E; a retrospective view is Boutros Boutros-Ghali, *Unvanquished: A U.S.–UN Saga* (New York: Random House, 1999), p. 27.

130. For contemporaneous perspectives, see Paul Lewis, "As the UN's Armies Grow, the Talk is of Preventing War," *New York Times*, March 1, 1992, sec. 4, p. 4, Ahmed Marcia, "A Secretary General on the Firing Line," *World Press Review*, October 1992, pp. 12–13; and Boutros Boutros-Ghali, "Beleaguered are the Peacekeepers," *New York Times*, October 30, 1994, p. A15.

131. E.g., George Will, "Bad Idea Whose Time Has Come at the UN," *Houston Chronicle*, April 12, 1993, p. 10A.

132. Ronald Reagan, "Evil Empire Gone, but Somalia Proves Evil Still Stalks Globe," *Houston Chronicle*, December 15, 1992, p. 11B. Reagan's 1992 statement

quoted in Bill Kaufman, *America First! Its History, Culture and Politics* (Amherst, N.Y.: Prometheus Books, 1995), p. 242.

133. Paul Lewis, "The Right to Intervene," *New York Times*, July 12, 1992, p. 22E; Christopher C. Joyner, "When Human Suffering Warrants Military Action," *Chronicle of Higher Education*, January 27, 1992, p. A52.

134. An essay on that contradiction is Henry Grunwald, "Memorandum to Woodrow Wilson," *Time*, November 14, 1994, p. 104.

135. Robert Oakley and Michael Dziedzic, "Policing the New World Disorder," *Strategic Forum* 84 (October 1996).

136. For a recent official vignette, see "The Bosnia Beat," *Peace Watch* 5:2 (February 1999): 6–7.

137. Olara A. Otunnu and Michael W. Doyle, eds., *Peacemakers and Peacekeeping for the New Century* (London: Rowman and Littlefield, 1998), p. 182.

138. A useful overview is Roxane D. Sismandis, *Police Functions: Report from a Workshop Organized by the United States Institute of Peace* [Peaceworks No. 14] (Washington, D.C.: United States Institute of Peace, 1997).

139. Robert M. McClure and Morton Orlon II, "Is the U.N. Peacekeeping Role in Eclipse?" *Parameters* 29:3 (autumn 1999): 97.

140. A lengthy critique is Earl H. Tilford Jr., *Halt Phase Strategy: New Wine in Old Skins . . . With Powerpoint* (Carlisle, Pa.: U.S. Army War College, 1998).

141. Steven Lee Myers, "Bomb. Missile. Bomb. Hey. It Looks Like a War," *New York Times*, February 21, 1999, sec. 4, p. 4.

142. Boutros-Ghali, *Unvanquished*, pp. 142–143.

143. Craig B. Whitney, "European Union to Form Common Military Force," *Houston Chronicle*, June 4, 1999, p. 33A.

144. E.g., George C. Wilson, "We're Sending Warriors When Cops Could Keep the Peace," *Army Times*, July 19, 1999, p. 54.

145. William Appleman Williams, *Empire as a Way of Life* (Oxford: Oxford University Press, 1980), p. 114; and Donald C. F. Daniel and Bradd C. Hayes, *Coercive Inducement and the Containment of International Crises* (Washington, D.C.: United States Institute of Peace Press, 1999).

146. A cautionary essay on those problems is Charles Maechling Jr., "Can't Haiti's Cops Just Fight Crime?" *New York Times*, November 29, 1994, p. A17.

147. A detailed analysis is William H. Lewis, "United Nations Role Sharing," *Strategic Forum* 83 (September 1996); also see Margaret P. Karns and Karen A. Mingst, "The Evolution of United Nations Peacekeeping and Peacekeeping Forces: Lessons from the Past and Challenge for the Future," in Michael T. Klare and Yogesh Chandrin, eds., *World Security: Challenge for a New Century*, 3rd ed. (New York: St. Martin's Press, 1948), pp. 200–228.

148. Matthew Cox, "Walking the Beat with KFOR's Warrior Cops,"*Army Times*, March 27, 2000, pp. 15–16.

149. Presidential Decision Directive-25 (PDD-25), signed on May 4, 1994. For a general lamentation regarding the surge and decline of UN peace keeping in the early 1990s, see Brian Urquhart, "Who Can Police the World?" *New York Review of Books*, May 12, 1994, pp. 29–33.

150. For details, see "Agreement on International Police Task Force," in *General Framework Agreement: Proximity Peace Talks Wright-Patterson Air Force Base, Dayton, Ohio*, Annex 11.

151. Elaine Sciolino, "Soldiering On, Without an Enemy,"*New York Times*, October 29, 1995, sec. 4, p. 1.

152. Barbara Crossette, "U.S. Politicians Less Supportive of UN than Public," *Houston Chronicle*, September 23, 1996, p. 9A.

153. "House Stymies Payment of Debt to UN," *Houston Chronicle*, November 14, 1997.

154. The trend left a faint trace in the media, e.g., "Who Are the UN Peacekeepers?" *Parade*, January 12, 1997, p. 18, and "United Nations' Peacekeeping Force Dwindles," *Parade Magazine*, October 12, 1997, p. 14.

155. E.g., see the "Sovereignty and International Law" section, *Harvard International Review* 27:3 (summer 1995): 8–39.

156. See Barbara Crossette, "UN Falters in Post-Cold-War Peacekeeping, but Sees Role as Essential," *New York Times*, December 5, 1994.

157. Myron H. Nordquist, *What Color Helmet? Reforming Security Council Peacekeeping Mandates* (Newport, R.I.: Naval War College, 1997) [Newport Paper No. 12], p. 66.

158. John C. Henry, "Ex-Army Medic Says Court-Martial Unfair," *Houston Chronicle*, February 5, 2000, p. 31A.

159. Brian Urquhart, "Whose Fight Is It?" *New York Times*, May 22, 1994, sec. 1, p. 15, and H. Francis Shattuck, Letter to *New York Times*, November 14, 1994, p. A10.

160. Most notably, Patrick J. Buchanan. E.g., see his "Bihac Sends Clear Signal to Put America First," *Houston Chronicle*, November 29, 1994, p. 17A, and "U.S. Needs to Put America First," *Bryan–College Station Eagle*, February 11, 1998, p. A10.

161. E.g., James A. Baker III, "Definite 'Do's and Don'ts' in U.S.–UN Relationship," *Houston Chronicle*, February 16, 1997, p. 4C, which overlooked such cases as the Boxer affair of 1900, the Western Front and Italy 1917–1918, the Archangel expedition of 1918–1919, the Normandy invasion and campaign in northwest Europe, 1944–1945, and NATO; for critiques of resurgent unilateralism and its implications for foreign relations and command-and-control, see Arthur Schlesinger Jr., "New isolationists Weaken America," *New York Times*, June 11, 1995, sec. 4, p. 3, and Claudia Dreifus' interview with General John Shalikashvili, "Who's the Enemy Now?" *New York Times Magazine*, May 21, 1995, p. 35.

162. Patrick J. Buchanan, *A Republic, Not an Empire: Reclaiming America's Destiny* (Washington, D.C.: Regnery Publishing Co., 1999).

163. Quoted in Fritz Lanham, "The Limits of Idealism," *Houston Chronicle, Zest* section, May 8, 1994, p. 27.

164. "Joint Maneuvers," *World Press Review*, May 1996, p. 23.

165. E.g., see "Helms' 'Hand of Friendship' Dishes Out Tough Talk to U.N.," *Houston Chronicle*, January 21, 2000, p. 28A.

166. In his *Conflict and Defense: A General Theory*, quoted in J. G. Starke, *An Introduction to the Science of Peace [Irenology]* (Leyden, Netherlands: A. W. Sijthoff, 1968), p. 108.

167. E.g., see Stephen Badsey, "The Media and UN 'Peacekeeping' Since the Gulf War," *Journal of Conflict Studies* 27:1 (spring 1997): 7–27.

168. Quotes from Richard N. Haas' *The Reluctant Sheriff*, in "Viewpoint," *Aviation Week and Space Technology* 147:10 (September 8, 1997): 70.

169. E.g., the favorable critique of the Clark–Sohn Plan in Revelle Heater, *World Citizenship and Government: Cosmopolitan Ideas in the History of Western Political Thought* (Houndsmills, England: Macmillan, 1996), pp. 149–154.

170. Tony Mason, "Operations in Search of a Title: Air Power in Operations Other Than War," in Richard Hallion, ed., *Air Power Confronts an Unstable World* (London: Brassey's, 1997), p. 173.

171. Myers, "Bomb. Missile. Bomb."

172. White, *Keeping the Peace*, p. 128; Pirnie and Simons, *Soldiers for Peace*, pp. 70–71.

173. "Enlarged Police Expedition Urged for Bosnia," *Houston Chronicle*, December 3, 1997, p. 32A; critical commentaries are George F. Will, "Saddam Has Given Severe Blow to UN Idea," *Houston Chronicle*, November 11, 1997, p. 46A; and Elaine Sciolino, "It Turns Out That All Global Politics is Local," *New York Times*, December 7, 1997, sec. 4, p. 3. For a post–Serbian air campaign perspective, see David Callahan, "The Lesson of Kosovo," *The Washington Monthly*, July-August 1999, pp. 34–37.

174. E.g., see Howard Banks, "Not-so-smart Bombs," *Forbes*, May 17, 1999, pp. 188–191; Susan Estrich, "Kosovo is No TV Special—It's Good vs. Evil," *Houston Chronicle*, May 6, 1999, p. 44A, and Robert Novak, "NATO Commander's Belligerency Showing," *Houston Chronicle*, May 6, 1999, p. 44A.

175. For contrary positions by members of Congress on that issue, see "Should the U.S. Pay 'Back Dues' to the UN?" *American Legion Magazine* 147:1 (July 1999): 44.

176. For comments on the "new imperialism," see Tom Hayden, "NATO's New Imperialism," *Mother Jones*, November/December 1999, p. 12; for a range of perspectives throughout the 1990s, see Paul F. Diehl, *International Peacekeeping* (Baltimore: Johns Hopkins University Press, 1993); Fisas Vicenx, *Blue Geopolitics: The United Nations Reforms and the Future of the Blue Helmets* (London: Pluto Press, 1995); Margaret Cecchine Harrell and Robert Howe, *Beyond Traditional Peacekeeping* (New York: St. Martin's Press, 1995); Geoff Simmons, *UN Malaise: Power Problems and Realpolitik* (New York: St. Martin's Press, 1995); Steven R. Ratner, *The New UN Peacekeeping: Building Peace in the Lands of Conflict After the Cold War* (New York: St. Martin's Press, 1995); Trevor Findlay, ed., *Challenges for the New Peacekeepers* [SIPRI Research Report No. 12] (Oxford: Oxford University Press, 1996); James H. Allan, *Peacekeeping: Outspoken Observations by a Field Officer* (Westport, Conn.: Praeger, 1996); Michael C. Pugh, ed., *The UN, Peace and Force* (London: Frank Cass, 1997); Michael Wesley, *Casualties of the New World Order: The Causes of Failure in U.N. Missions to Civil Wars* (New York: St. Martin's Press, 1997); Jocelyn Coulon, *Soldiers of Diplomacy: The United Nations, Peacekeeping, and the New World Order* (Toronto: University of Toronto Press, 1998); Roger Williamson, ed., *Some Corner of a Foreign Field: Intervention and World Order* (New York: St. Martin's Press, 1998); Joseph P. Lorenz, *Peace Power and the U.N.: The Uses and Limits of Collective Force* (Boulder: Westview Press, 1999); and Claire Trean, "Reinforcing the United Nations," *Le Monde*, May 24, 2000, reprinted in *World Press Review*, August 2000, pp. 15–16. Perhaps the most lyrical and traditional, if not atavistic, is Ralph Peters, "The American Mission," *Parameters* 29:3 (autumn 1999): 2–8.

177. Critiques of the air campaign include Paul Mann, "Belgrade Called Victor

in War's First Phase," *Aviation Week and Space Technology* 150:17 (April 26, 1999): 28–30; Joel R. Rawson, "Wars are Won on the Ground," *Bryan-College Station Eagle,* May 5, 1999, p. A10; George Will, "Yet Another Demonstration of Air Power's Limits?" *Houston Chronicle,* March 28, 1999, p. 3C; for an overview of financial costs, see Eric Schmitt, "It Costs a Lot to Kill More People," *New York Times,* May 2, 1999, sec. 4, p. 5.

178. Steven Lee Myers, "All in Favor of This Target, Say Yes, Si, Oui, Ja," *New York Times,* April 25, 1999, sec. 4, p. 4; and Niall Ferguson, "NATO Celebration Belies Europe's Mistrust of U.S.," *Houston Chronicle,* April 23, 1999, p. 39A. For views on the official "spin," see Pierre Sparaco, "Allied Teamwork Deemed Effective," *Aviation Week and Space Technology* 150:16 (April 19, 1999): 30, and "Security Concerns Impede Alliances," *Aviation Week and Space Technology* 150:17 (April 26, 1999): 37.

179. Paul Mann, "Kosovo's Lessons Called Ambiguous," *Aviation Week and Space Technology* 150:25 (June 28, 1999): 32–33.

180. E.g., see David Reynolds, *One World Divisible: A Global History Since 1945* (New York: W. W. Norton, 2000).

181. E.g., see Judith Miller, "Security Council Relegated to Sidelines," *New York Times,* March 14, 1999, sec. 4, p. 14, and John B. Alexander, *Future War: Non-Lethal Weapons in Twenty–First Century Warfare* (New York: St. Martin's Press, 1999), p. 29.

182. I am grateful to Colonel Ron Ladnier USAF, of the Political and Military Affairs Branch, Department of State, for his insights on these matters.

183. Ian O. Lesser, Bruce Hoffman, John Arquilla, David Ronfeldt, and Michael Zanini, *Counter the New Terrorism* (Santa Monica, Calif.: Rand Corporation, 1999), pp. 75, 136; all the NATO allies involved, except Britain, refused to provide the U.S. Air Force with weapons accuracy data; see "State Secrets," *Aviation Week and Space Technology* 152:10 (March 6, 2000): 23.

184. "NATO Toll: 500 Civilians?" *Houston Chronicle,* February 7, 2000, p. 14A.

185. A recent example is Randall Rice, "Bombing Auschwitz: U.S. 15th Air Force and the Military Aspects of a Possible Attack," *War in History* 6:2 (April 1999): 205–229.

Index

Addams, Jane, 12; role in creating League to Enforce Peace, 9

Advisory Committee in Post-War Foreign Policy. *See* United States: Department of State

Afghanistan, 35; British bombing of, 38; Soviets depart, 1989, 152; Soviets invade, 1979, 151

AFHQ (Air Force Headquarters), created, 1935, 86

African troops, in the Congo, 142

Air control. *See* Air policing

Aircraft, media imagery of, 86

Aircraft types:

- B-17 *Flying Fortress*, 87; interception of *Rex*, 51
- B-24 *Liberator*, 87
- B-29 *Superfortress*, 97, 137; based in China and Marianas, 109; Cold War role, 132 ; in Korean War, 134; in SILVERPLATE, 104
- B-36 *Peacemaker*, 87, 132; not used in combat, 150; role in "Battle of the Potomac," 133
- B-47 *Stratoet*, not used in combat, 150
- B-50, in Cold War, 132; in Korean War, 134
- B-52 *Stratofortress*, bomb North Vietnam, 149; tactical bombing by, in South Vietnam, 146, 148. *See* also ARCLIGHT
- B-58 *Hustler*, not used in combat, 150

Air Force magazine, UN space warning patrol proposed in, 1960, 144

Air policing: in Balkans, 153–156; in empires after World War I, 20; as "Euphemism for Empire" in Cold War, 131; and Geneva arms talks, 41–43, 48; IAF advocates cite as precedent, 39; as peace-keeping model, 102. *See also* Air control; Churchill, Prime Minister Sir Winston S.; Project CONTROL; Royal Air Force (RAF); World Disarmament Conference, 1933–1937

Air power, as means of "policing," 137–138; in Balkans, 1990s, 153–158; in Cold War, 123, 131, 140; enthusiasts promote, 51; enthusiasts' goals realized, 1945, 112;

feared as political coercion tool, 1930s, 40, 42; Franklin D. Roosevelt's enthusiasm for, 56; in Gulf War, 154–155; in Korean War, 133–35; limits of recognized, in Korean War, 136; limits of recognized, in low-intensity conflict, 141; limits of recognized, in 1930s, 56, 58; limits of recognized, in Vietnam War, 148; Lippmann's paean to, 103; as neo-imperialism, 158; role of, in warfare, debated, 37, 43; Slessor's views on limits of, 109; symbolic use of, 1920s-1930s, 42; U.S. Navy accepts utility of, 85; in Vietnam War, 146, 147; in World War I, 11, 19. IA; *See also* Aircraft types; air policing; Royal Air Force (RAF); Trenchard, General/Air Chief Marshal Lord Hugh; United States: Air Force; United States: Army Air Corps; United States: Army Air Forces; United States: Navy; World War I; World War II

Air War Study Group, Cornell University, 147

A League of Airmen (RAND Study), 152

Allenby, Field Marshal Lord, proposes IAF, 1936, 53

Allies/Allied Powers, in World War I, 23–25; vengefulness of, at Paris Peace Conference, 103

American Farm Labor Federation, Lord Lothian speaks to, 1940, 74

American Legion, oppose international super state, 94–95; magazine urges creating "world police force," 152

American South, fears of IPF in, 78, 143–44

Angell, Norman, 13; changes view on IPF, 1930s, 58; quoted on League with "teeth," 102

Anglo-American air offensive, Europe, WW II. *See* Combined Bomber Offensive

Anglo-American alliance, Wilson threatens France with, 1919, 26

Anglo-American naval peace-keeping patrol: in post–World War II plans, 102, 145; proposed in *Popular Science*, 1944, 101–102; Teddy Roosevelt proposes, 1910, 8–9, 56, 78

Anglo-American relations: during Cold War, 138; "special relationship," 74–75, 103, 138; during WW II, 85, 88, 89, 101, 125

Appeasement, by Western democracies, of Germany, Italy, and Japan, 1930s, 33, 40, 54, 62

Arbitration. *See* Peace arbitrations

ARCLIGHT, tactical bombing by B-52s in Vietnam War, 146

Area bombing, by RAF and USAAF, 109

Argentina, 14; U.S. bombers fly to, 51

Argentine Air Force, uses Exocet missiles, 1982, 150

Arms control talks, 142. *See also* London Naval Conference; Washington Naval Conference; World Disarmament Conference

Army Air Corps, U.S. *See* United States: Army Air Corps.

Army Air Forces. *See* United States: Army Air Forces

Arnold, General of the Army Henry H. "Hap," foresees UN failure, 127; heads Army Air Corps and AAF, 52; and nuclear weapons, 112; ponders air power future, 1946, 113; seeks independent air force, 86, 90. *See also* Combined Chiefs of Staff; United States: Army Air Corps; United States: Army Air Forces; United States: Joint Chiefs of Staff

"As Easy as A.B.C." (Kipling), 9

Atlantic Charter, 1941, 77, 80, 96; leads to United Nations, 83

Atmospheric Test Ban Treaty, 1963, 139
"Atomic cannon," 132; in Korean War nuclear diplomacy, 142
Attlee, Clement. *See* Lord Attlee
Australia/Australians, in WW II, 92; in Vietnam War, 147
Austria, Hitler seizes, 40
Aviation technology: evolution of, in World War I, 19; between world wars, 37, 48, 51, 52, 56, 73
Axis powers: alliance formed to counter, 161; rise of, late 1930s, 40; triumph of expected, 61

Baldwin, Prime Minister Stanley: defines deterrence, 49; government of, berated by Churchill, 58; supports sanctions, 49
Balkans, 72; U.S. intervention in, 1990s, 153–156. *See also* Bosnia; Kosovo; Operation DELIBERATE FORCE; Serbia; United Nations; United States: Air Force
Ball, Senator Joseph, proposes world government system, 95
Ball-Burton-Hatch-Hill Resolution, U.S. Congress (B2H2), 1943, 95
Baruch, Ambassador Bernard, nuclear disarmament plan of, 124
Battle of the Atlantic, 92
Battle of Britain, 1940–1941, 72, 96; Czechs and Poles in, 76
Battle of the Bulge, 1944, 110
Battle of the Coral Sea, 1942, 89, 92
Battle of El Alamein, 1942, 92
Battle of Germany, 1944–1945, 109
Battle of Khafji, 1991, 155
Battle of Midway, 1942, 89
"Battle of the Potomac," 1948–1949, 132, 140
Battle of the Solomons, 1942–1943, 89
Becker, Carl: proposes "League with teeth," 99; supports British imperialism, 79
Belgium, 25, 109, 144; withdraws from Congo, 142
Bellers, John, plan for European Union, 3–4
Benes, Edouard, opposes IAF, 42
Bennett, Air Vice Marshal Don, 57, 100
Berbers, French use of air power against, 38
Berlin Airlift, 1949–1950, 132, 133
Berlin Crisis, 1961, 145
"Big Five" (Britain, China, France, U.S., USSR), 101, 108
"Big Four" (Chiang Kai-shek, Churchill, Roosevelt, Stalin), 103; issue joint statements, 96; Roosevelt proclaims unity of, 111
"Big Four" (Clemenceau, Lloyd-George, Orlando, Wilson), at Paris Peace Conference, 1919, 21, 24
"Big Three" (Churchill, Roosevelt, Stalin): divisions among, 84, 85, 110. *See also* Teheran
Bikini atoll, atomic tests at, 1946, 131, 132
"Blitz," German air campaign versus Britain, 1940–1941, 72, 73. *See also* Battle of Britain.
Blitzkrieg, 71, 72
Blockade, of Central powers by Allied powers, 25
Blum, Léon: challenge by Hans Wehrberg, 45; opposes international armed force before disarming, 42
Boeing Aircraft Co., builds B-17, 51. *See also* Aircraft types
Bolshevism, fears of, 36, 54
Bomb damage assessment, in Vietnam War, 147; in World War II, 76–77
"Bomber Barons," U.S. and British, 76, 96
Bomber Command, RAF, 76, 89; halts raids on Germany, 1943, 97
Bomber Command, U.S. Air Force, IAF role of planned, 132
Bombers, attempts to ban, 46–47; fears of, 1930s, 55; media images of, 1930s, 50

Bombing accuracy, overestimated by air planners, 56, 76
Bonapartism, aerial, fears of, 1930s, 42, 60
Bonnet, Henri, proposes regional collective security, IAF, 100
Borah, Senator William: opposes League of Nations Covenant, 28; proposes outlawing war, 39
Bosnia, UN–NATO interventions in, 154
Boulding, Kenneth, proposes "Open Skies" flights, 157
Bourgeois, Léon: IPF proposals of, 60; predicts League demise, 26, 47
Boutros-Ghali, UN Secretary General Boutros, proposes on-call national UN contingents, 153
Braham, Harry Lewis, quoted, on causes of League's failure, 99
Brigadistas (International Brigade members) seen as IPF in Spanish Civil War, 57
Bright, John, skepticism of using force to maintain peace, 4–5
Britain. *See* Great Britain
British Commonwealth Division (Korea, 1950–1953), 134
British Commonwealth of Nations: created by Statute of Westminster, 46; as "experiment in democracy" and "League of Nations," 38, 46; as one of "four policemen," 81; troops in Korean War, 134
British Empire, 78, 101; decline of, 131; philosophers praise, 103; use of air power in, 20, 35, 38, 41–43, 48
British League of Nations Union, sponsors air conference, 52
British Strategic Bombing Survey, weakens strategic bombing rationale, 129
Brodie, Bernard: develops nuclear doctrine, 132; identifies nuclear weapons dilemmas, 127, 139
Brookings Institution, 1943 white paper, on postwar world order, 103; deems IPF too costly, 104
Brushfire war, 141, 157. *See also* Counterinsurgency (COIN); Limited wars; Unconventional warfare
Bunche, Ambassador Ralph, supports "UN Peace Force," 128
Burton, John, on unenforceability of peace, and dangers of peacekeeping forces, 141
Bush, President George H. W., and Gulf War, 153, 158
Butler, Nicholas Murray, calls for "combined navies," 101
Bryan, William Jennings, as Wilson's secretary of state, 22
Byrnes, Justice/Secretary of State James F., blames Soviets for UN powerlessness, 126–127

Cadogan, Sir Alexander, at Dumbarton Oaks, supports Stettinius against IAF at Dumbarton Oaks, 108
Canada: boundary disputes, arbitration of, 6; configures armed forces for UN role, 126. *See also* King, Prime Minister Mackenzie; Pearson, Lester
Carr, E. H.: quoted on fragility of 1919 peace, 41; rejects IPF concept, 60
Casablanca Conference, 1943, 82, 90, 93
Casualty projections: from air attack, in Munich crisis, 72; in World War II, 73
Cecil, Robert. *See* Lord Cecil
Century Club conferences, and League to Enforce Peace, 9
Chamberlain, Prime Minister Neville: government of, berated by Churchill, 58; and Munich crisis, 55, 72
Charter, United Nations. *See* United Nations Charter

Chiang Kai-shek, 83, 87; distrusted by Churchill and Stalin, 96
Chiang Kai-shek, Madame, supports "four policemen" postwar peace enforcement model, 1942, 81
Chicago Bridge speech, 1937. *See* "Quarantine the aggressors" speech
Chicago Council on Foreign Relations, Lord Lothian's speech to, 1940, 78
Chicago International Aviation Conference, 1944, 48, 106
Chiefs of Staff (British), favor IPF/IAF, 108
China, 52, 72, 78, 80, 108, 125; delegates leak Dumbarton Oaks proceedings, 106; favors standing UN IAF at Dumbarton Oaks, 107; Japanese bombing of, 47, 49, 59, 72, 96; Japanese claims in, 1919, 26; Maoist victory in, 1949, 131; media images of bombing in, 1930s, 44; Nationalist forces of, in World War II, 78; as one of "four policemen," 81, 112; U.S. garrisons in, 35; U.S. supports "Big Four" membership, 100. *See also* Chiang Kai-shek; Shanghai
China, People's Republic of: air forces of, in Korean War, 135; clashes with USSR, 138; develops nuclear weapons, 142; enters Korean War, 135; rapprochement with USSR, 133
Chinese Civil War, 1946–1949, 141
Churchill, Prime Minister Sir Winston S., 50, 55, 84, 92, 110, 125; as architect of "air control," 38; critiques American isolationism, 34; distrusts Chiang Kai-shek, 81, 96; distrusts de Gaulle, 81; favors limiting sovereignty, 45, 94; Guildhall speech, 1943, 93; Iron Curtain speech of, 1946, 125; on London's vulnerability to air attack, 35; Moscow meeting with Stalin, 1942, 93; opposes aid to France in Indochina, 1954, 138; opposes appeasement, 45, 58; quoted, on postwar world organization, 100, 102; speaks to Congress, 1943, 93; supports IAF/IPF proposals, 45, 82, 111; urges Anglo-American postwar collaboration, 91; voted out of power, 1945, 103, 112. *See also* Air policing; Casablanca Conference, 1943; Sovereignty; World War II
Clapper, Raymond, praises Henry Wallace's "common man" speech, 80
Clark, Grenville, proposes "World Police Force" after Vietnam, 149; IPF design of, with Louis Sohn, 128
Clark-Sohn plan, 128; in 1990s, 157; revived in 1961, 145. *See also* McCloy-Zorin "Joint Statement of Agreed Principles for Disarmament Negotiations"
Clinton, President William Jefferson, sets peace-keeping guidelines, 1994, 156
Clinton administration: backs away from UN, 157; reverses course on intervention, 153, 158
COIN. *See* Counterinsurgency
Cold War, 113, 133, 138, 139, 157; air power in, 123; decolonization during, 128; nuclear weapons and, 136; patterns of conflict in, 123; undercuts plans for UN IPF/IAF, 125
Collateral damage, 151, 152
Collective security, 35, 40, 44, 45, 49, 52, 57, 58, 61–62, 98; critiques of, during Cold War, 136; critiques during World War II, 95; fades during Cold War, 124, 127, 145; as *guerre totale*, 73; nuclear weapons effect on, 112, 124, 136; opposed by Nixon and Kissinger, 150; pacifists debate utility of, 1930s, 56; religious proposals for, 75; resurgence of, 1990s, 158;

Roosevelt vague on, 77; Wallace and Welles modify views on, 91. *See also* League of Nations; United Nations
Combined Bomber Offensive (U.S.–British), 92; suspended, 97
Combined Chiefs of Staff (U.S.–British), 125; approve IPF, 97; created, 1942, 88
Comic strips, IAF/IPF portrayals in, 75–76
Commission to Study the Organization of Peace. *See* United States: Department of State
Committee to Defend America by Aiding the Allies, 75
Communist bloc: propaganda of, 138, 141; rejects non-Soviet world order concepts, 129; tensions within, 138. *See also* China, People's Republic of; Russia; Soviet Union; USSR
Concert of Europe, 4
Conference on Security and Cooperation in Europe (CSCE), 154
Congo, 144; UN troops deployed in, 142
Congress, U.S. *See* United States: Congress
Congress of Vienna, 1815, 4
Connally, Senator Tom, resolution of, 1943, 95
Conservative Party (British), opposes Geneva Protocol, 37, 49
Convention of Paris (aviation), 1919, 106
Coolidge, President Calvin, U.S. president, disarmament plans thwarted, 35
Corn, Joseph, and aviation as a cult, 19
Cornell University, Air War Study Group, 147
Coudenhove-Kalergi, Count Richard Nicholas, proposes Pax Americana, 99–100
Council on Foreign Relations, 36, 74
Counterinsurgency (COIN), 140, 142; theories of, 141. *See also* Brushfire war; Limited wars; Unconventional warfare
Covenant of the League of Nations, 9, 12, 14, 21–28, 34, 37, 47, 50, 52, 74, 109; revisions proposed, 53; on use of force, 109; U.S. Senate endorses principles of, 1935, 51; U.S. Senate rejects, 26–28, 50, 98. *See also* League of Nations
Crucee, Emeric, 3
Cruise missiles, 150; used against Serbia, 1999, 155
Culbertson, Ely, IPF plans of, 91; Dorothy Thompson critiques, 94
Czechoslovakia, and Munich crisis, 40, 45, 50, 54, 61; Communist coup in, 1948, 13

Daladier, Premier Edouard, and Munich crisis, 55
Dante, 12; proposes worldwide superempire, 2, 126
Dayton peace talks, 1995, International Police Task Force (IPTF) ground rules set at, 156
Decolonization, during Cold War, 128
de Gaulle, General/President Charles, 52; Churchill distrusts, 81
de la Gresseria, Raoul, international army scheme of, 6
Demobilization, of U.S. armed forces, 1945–1946, 113, 130
Department of State, U.S. *See* United States: Department of State
de Saint Pierre, Abbé, peace plan of, 4, 8
Desert Air Force, as type of IAF, 76
DESERT STORM. *See* Operation DESERT STORM
de Seversky, Alexander P.: claims air force power monopoly, 129; promotes strategic bombing, early World War II, 72; urges independent air force, 91
"Destroyer deal," Anglo-American, 1940, 75; validates imperialism, 101

de Sully, Duc, 2, 8
Dewey, Governor Thomas E., as internationalist Republican, 98
Dickinson, G. Lowes, proposes LTEP, 10, 12; deems America unique at 1919 Peace Conference, 21
Dictators, 1920s-1930s, 40, 53; brutality of, 39; Roosevelt opposes covertly, 56; use of air power by, 44
Diem, Premier Ngo Dinh, regime of, falls, 1963, 146
Dienbienphu, Viet Minh defeat French at, 1954, 137, 138
Disaster Through Air Power (Andrews), 129
Disney, Walt, produces *Victory Through Air Power*, 91
Doenitz, Grand Admiral Karl, 92
Douglas, Senator Paul, deems Korean War UN police action, 133
Douhet, Giulio, 33, 43
DPKO. *See* United Nations: Department of Peacekeeping Operations
du Bois, Pierre, proposes European order system, 2
Dulles, Secretary of State John Foster: deems nuclear weapons a "wasting asset," 139; massive retaliation policy of, 136
Dumbarton Oaks, United Nations Organization Planning Conference at, 1944, 80, 81, 91, 94, 98, 102, 104, 109–111; imperialism and, 101; sovereignty at, 106; Stalin approves outcomes of, 111; U.S. armed forces' clashes over, 101–102, 113. *See also* China; France; Great Britain; International air force concept; International police force concept; Special Informal Military Group (SIMG); Stettinius, Undersecretary of State Edward M.; United Nations; United Nations Charter; USSR

Eagle Squadron, U.S. volunteers in RAF, 76
Earle, Edward Mead, Cold War defense model proposed by, 125
Eberstadt, Ferdinand, authors antiunification study, 130
Eden, Foreign Minister Anthony, later prime minister: compares Nasser to Hitler, 158; in Saar plebiscite, 46; supports FDR's "four policemen" postwar peace enforcement model, 1942, 81. *See also* Suez affair
Egypt, in Suez affair, 142
Eichelberger, Clark, helps draft UN Charter, 95
8th Air Force. *See* United States: Army Air Forces
Einstein, Albert, favors collective security, 57
Eisenhower, General of the Army/President Dwight D., 85; condemns interservice bickering, 1949, 133; considers aiding French in Indochina, 1954, 137; makes gentlemen's agreement with Spaatz, 1946, 131
Eisenhower administration: conflicting collective security policies of, 137; ends Korean War, 136; massive retaliation policy of, 136, 138; New Look defense policy of, 136; nuclear diplomacy of, 142; proposes "Open Skies" flights, 157; supports UN constabulary, 141. *See also* Dulles, Secretary of State John Foster; "Lucky Dragon" incident; Massive retaliation policy; "New Look"; Suez affair
Eliot, Major George Fielding, proposes UN IAF/IPF in *Look*, 111; postwar "four policemen" model of, 125
Elizabeth I, as architect of *Grand dessein*, 2–3
Epictetus, proposes world order system, 1
Erzberger, Matthias, proposes international peace-keeping contingents, 27

Escalation theory, in Vietnam War air war, 147
Ethiopia, 47, 52; armed forces, in World War II, 78; bombing of, in mid-1930s, 44; Italian invasion of, 40, 53, 58
Eurasia, "rimlands" of, in Cold War, 138
European air force proposals, 59, 136
European Union, creates military forces, 155
Executive Council, Body of Delegates (U.S.), 23

Fabian Society, endorses collective security proposals, 10
Falklands War, 1982, smart weapons in, 150
Far East Bomber Command. *See* United States: Armed Forces Far East
Far Eastern Air Force. *See* United States: Armed Forces Far East
Fascism/Fascists, 42, 76; air power of, 55; aviation used as modernist symbol by, 37; influence in Latin America, 52. *See also* Italy; Mussolini, Benito
Federal marshals (U.S.), as IPF model, 151
Finletter, Secretary of the Air Force Thomas K., opposes expanding UN authority, 137
Force de frappe, French nuclear bomber force, 138
Forrestal, Secretary of the Navy James V., as Secretary of Defense, 130, 133; death of, 132; opposes separate air force, 88; opposes unification, 130. *See also* "Battle of the Potomac"; National Security Act of 1947; "Revolt of the Admirals"
"Fortress America" concept, proposed by Herbert Hoover, 52, 72, 74
Fosdick, Harry Emerson, on "paganism" of nationalism, 7
"Four Freedoms": FDR speech, 1941, 77; unevenness of, in Allied camp, 80–81
"Four policemen," postwar peace enforcement model proposed by FDR, 81, 84, 91, 110; Roosevelt offers to Stalin at Teheran, 96; Soviets support at Dumbarton Oaks, 105. *See also* International police force concept; Roosevelt, President Franklin D.
Fourteen Points, 10, 14, 20, 21, 23, 24, 27, 34; contrasted with Atlantic Charter, 77
France, 104, 109, 110; accepts FDR's bombing limits, 1939–1940, 71; Anglo-American suspicion of motives, 25, 45; appeases dictators, 1930s, 44, 47; blamed for League's failure and World War II, 61, 72; Britain supports "Big Five" membership of, 100; British distrust interwar diplomacy of, 60; bullied by Hitler, 1938, 40; creates *force de frappe*, 138; defeated at Dienbienphu, 137; divergent defense contingencies of, 1930s, 57; fall of, 1940, 87; gains "Big Five" status, 81, 101; as imperial power, 14; interwar support of IPF, 26, 41, 42; and Munich crisis, 55; offers UN troops, 1992, 154; opposes Britain over Germany, 1922, 35; policy failure in Greco-Turkish War, 1922, 35; post–World War II constitution of, limits sovereignty, 126; proposes "Western Air Pact," 52; rearms, late 1930s, 56; reduces overseas involvement, 1950s-1960s, 138; in Suez affair, 142; supports strong League of Nations, 23–27; views on Allied strategic bombing force, World War I, 11; Wehrmacht offensive in, 1940, 71. *See also* Blum, Léon; Bourgeois, Léon; Clemenceau, Georges; de Gaulle, General/

President Charles; Dumbarton Oaks, United Nations Organization Planning Conference at; French Foreign Legion; Herriot, Edouard; international air force concept; International police force concept; Kellogg-Briand Peace Pact; League of Nations; Paris Peace Conference; Paul-Boncour, Josef; Suez affair

Franco, General Francisco, 50, 61. *See also* Spanish Civil War

Frederick the Great, rejects armed peace enforcement, 22, 28

Freedom from War (Bennett), calls for IAF, 100

French Empire, 78; air power use in, 38; decline of, 131

French Foreign Legion, as IPF model, 57, 59, 76, 142

French generals: and Maginot Line, 39; oppose League Covenant, 22

French Indochina, war in, 1948–1954, 140. *See also* Dienbienphu; France

Fulbright, Senator J. William, 108; calls for peace-keeping system, 1943, 95

Fulljames, Wing Commander R., proposes multinational UN air contingents, 127

Gallup Poll, 72–73

General staff: European Union, 155; Prussian, 25. *See also* International general staff

Geneva, Switzerland, 4, 41, 149; arms control talks at, 1927, 35; arms control talks at, 1932–1937, 41–42; as League of Nations headquarters, 34

Geneva Protocol, 1924, 37, 50

Geopolitics, 84; popularity of, during WW II, 83, 103, 111

German Army, fights on, 1944, 109; officers urge UN air force, 127

Germany, 49, 104; accepts FDR bombing limits, 1939–1940, 71; as Axis power, 40; bombs Britain, 11; colonies of, 24; diplomats oppose IAF, 42; Franco-British differences over, 1922, 35; granted equal rights, 1932, 48; industry damaged by Allied bombing, 97; leaves League, 50, 52. *See also* Hitler, Adolf; Luftwaffe; Treaty of Versailles; Wehrmacht; Weimar Germany/Republic; World War I; World War II

Gibson, Wing Commander Guy, 57, 100

Grand dessein, of Henri IV, 2–3

Gran Chaco War, 47, 49, 52

"Grand Police Alliance," 1919, concept rejected by Wilson, 23

Great Britain, 78, 154; accepts FDR bombing limits, 1939–1940, 71; appeases dictators, 1930s, 44; blamed for League's failure and World War II, 61, 72; breaks alliance with Japan, 36; bullied by Hitler, 1938, 40; Churchill voted out, 1945, 103; decline of empire, 131; divergent defense contingencies of, 1930s, 57; faces Nazis and Fascists alone, 76; favors IAF at Dumbarton Oaks opening, 105, 107, 108; and Greco-Turkish War, 1922, 35; as imperial power, 14, 36, 41; and Munich crisis, 55; nuclear age vulnerability of, 125; opposes France, 35, 36; Oxford Union Debate, 1938, 58; Peace Ballot in, 1938, 58; power of, in decline, 104; proposes Allied strategic bombing force, 1917, 11; proposes European air force, 136; proposes "Western Air Pact," 52; rearms, late 1930s, 56; reduces overseas involvement, 1950s-1960s, 138; remains outside Concert of Europe, 4; Saar police contingent of, 46; signs Atmospheric Test Ban Treaty, 139; students in, oppose conscription, late 1930s, 39; in Suez affair, 142; suspects French diplomacy, 41–

42, 60; in UN armed contingent, 112; War Cabinet of, orders Berlin bombing, 1940, 72; wins Schneider Trophy races, 42. *See also* Air policing; British Commonwealth of Nations; British Empire; Churchill, Prime Minister Sir Winston S.; Eden, Foreign Minister Anthony; House of Commons (British); Leftists, Great Britain; Rightists, Great Britain; Royal Air Force (RAF); Royal Navy; Trenchard, General/Air Chief Marshal Lord Hugh; World War I; World War II

Great War, 1914–1918, 19, 20, 33. *See also* World War I

Greco-Turkish War, 1922, British-French policy failure in, 35

Greece, 34, 111, 112, 141

Grotius, Hugo, peace union proposals of, 3

Group to Study the Organization of the Peace. *See* United States: Department of State

Gulf of Tonkin affair/Resolution, 1964, 146

Gulf War, 1991, 34, 76, 151, 153; approximation of International Air Force in, 152; Battle of Khafji, 155; carbon-filament bombs used in, 152; debate over air power effect in, 154–155; UN involvement in, 152. *See also* Hussein, Saddam; Kuwait; Operation DESERT STORM

Gunboat diplomacy, 6, 140

The Hague, peace conferences at: nineteenth century, 7; 1915, 10

Haiti, UN intervention, 1990s, 153, 155

Halsey, Admiral of the Fleet William F., "Bull," 89, 130

Halt Phase Strategy/Doctrine, 154

Hammerskjold, UN Secretary General Dag:" deploys UN troops in Congo, 142; killed, 144

Hankey, Sir Maurice: distrusts Americans, 36, 61; urges British unilateralism, 35

Hart, Admiral Thomas, deems IPF unlikely, 1943, 97

Heilbrunn, Otto, on nuclear weapons and Cold War conflicts, 124

Hemispheric Defense, U.S. defense policy of, 51, 52, 72, 74

Henderson, Sir Arthur, proposes League IPF, 53

Henri IV, *Grand dessein* of, 2–3, 158

Herriot, Edouard, IPF proposals of, 41–42, 48, 50, 52, 58, 60, 149. *See also* World Disarmament Conference, 1933–1937

Hiroshima, atomic bombing of, 1945, 112, 113, 124, 132; as Communist propaganda theme in Cold War, 138

Hitchcock, Senator Gilbert, strongly supports ratifying League of Nations Covenant, 28

Hitler, Adolf, 33, 45, 51, 75; becomes chancellor and Fuehrer, 40, 46, 47, 48; breaks Versailles Treaty, 47; cited as precedent in post-1945 interventions, 158; creates Luftwaffe, 55; leaves League of Nations, 47; and Munich crisis, 55, 61; rise and aggressions of, 40, 47. *See also* Germany; Luftwaffe; Nazis; Rhineland; World War I; World War II

Holland, 109; as imperial power, 14; Saar police contingent of, 46

Holt, Hamilton, and LTEP formation, 12

Holy Alliance, suppresses liberalism, 4

Hoover, President Herbert: proposes "Fortress America" concept, 52, 72; seeks bomber ban, 46, 48, 49; supports a new League with "teeth," 84

Hopkins, Pryns, supports IPF concept, 59

House of Representatives (U.S.). *See*

United States: House of Representatives
Hughes, Charles Evans, 10; defines IPF paradox, 61–62
Hugo, Victor, United States of Europe plan of, 6
Hull, Secretary of State Cordell, 84, 91, 93, 111, 112; clashes with Welles, 90; creates postwar study groups, 73, 79, 80, 81; favors "Security Club," 78; gives Stettinius free hand, 112; rejects "superstate," 106; suspends planning groups, mid-1943, 94. *See also* International police force concept; United States: Navy; World War II
Hussein, Saddam: alliance formed to counter, 161; invades Kuwait, 1990, 152; remains in power, 153; violates Gulf War cease-fire, 154. *See also* Gulf War; Iraq; Kuwait; Operation DESERT STORM
Huxley, Julian, proposes "Security Club," 78
Hyde Park agreement, 1944, 125

IAF. *See* International Air Force concept
ICBMs. *See* Intercontinental ballistic missiles
IFOR. *See* United Nations Implementation Force
Imperialism, 14, 46, 101, 103; air power as a form of, 20, 41–43, 48, 157; American views on, WW II, 79, 101, 102; basing system of, 111; British Labourites oppose, 1930s, 57; in Cold War, 124; declines during Cold War, 128, 131, 140; eroded by air power, 84; forms of, in nineteenth century, 6; in tension with mother country democracies, 36–38; and Versailles treaty, 19, 21. *See also* British Commonwealth of Nations; British Empire; France
Independent Air Force, 43; cited as precedent by IAF advocates, 39, 59; formed in 1918, 11; mixed views of, 12
India: Britain leaves, 131; northwest frontier, air policing of, 20; troops in the Congo, 148
Insurgencies, in Cold War, 138, 141; fire-fighting metaphor of, 140
Intercontinental ballistic missiles (ICBMs), 145, 150; Soviets test, 1957, 144
International air force concept, 104, 130, 147, 152, 157; approximation of, in Gulf War, 153–156; British Empire plans for, 35; complexities of, 109; critiques of, 44, 83–84, 99; Culbertson plans for, 91; designs of, 38–40, 44, 52–54, 56, 57, 78, 83, 90, 94, 96, 125–127, 151; French proposals for, 4; future prospects considered, 159, 160; Hoover supports, 84; ignored in 1944 U.S. election, 98; Independent Air Force as model of, 11; Kipling's visions of, 9, 60; in Korean War, 133–135; national contingent model of, 56; New Commonwealth promotes, 1930s, 58; pacifists oppose, 1920s-1930s, 39, 56; plans for proliferate, 1940s, 62, 92; portrayals of, in comic strips, 1940s, 75–76; post-Vietnam disinterest, 149, 158; public support for increases, in 1930s, 49, 50, 58, 59; public support for increases, during World War II, 73, 83, 90, 91, 99; revived in Balkans, 1990s, 154; Soviets support creation of, 1944, 107; standing model of, 56; tactical rationale eroded by Cold War, 129, 145; as tool of imperialism, 38; traces of, in Balkans air campaigns, 1990s, 158; in UN Charter draft, 95; USAF Bomber Command, in IAF role, 132; U.S. Army Air Force plans for, sidelined at Dumbarton Oaks, 105,

106–108; wanes during Cold War, 113, 140, 141, 144 , 145; Wallace supports, 77, 94; Wells' vision of, 42, 60; between world wars, 37, 46, 47, 58. *See also* Dumbarton Oaks, United Nations Organization Planning Conference at; International police force concept; League of Nations; Royal Air Force (RAF); United Nations; United States: Air Force; United States: Army Air Forces; World War II

International Air Force Corps, Soviets propose at Dumbarton Oaks, 107

International Army, Julian Huxley proposes, 78

International Aviation Conference. *See* Chicago International Aviation Conference

International Brigades (*brigadistas*), seen as IPF, 57. *See also* International police force concept; Spanish Civil War

International Conference for the Maintenance of Peace, 1937, 62

International general staff, 59; French propose at Paris, 1919, 25, 26; French propose for UN, 101; Military Staff Committee, UN, as a form of, 110

Internationalism/internationalists, 108, 111–112; influence on Stettinius, 111; public support for grows during WW II, 91; U.S., grassroots anxiety, 103; U.S., public support for grows, late 1930s, 72

International police force concept, 1, 11, 24, 38, 48, 56, 58, 61, 80, 104–115, 149, 152; Americans support, WW II, 73, 83, 90, 99; as tool of imperialism, 20, 38; Boutros-Ghali design of, 153; Clark-Sohn plan for, 128, 145; complexities of, 109; comic strips portray, 75–76; critiques of, 13, 44, 75, 94, 99, 104, 134, 141; Culbertson plans for, 91; designs of, 38–41, 44, 52, 57, 74, 75, 80, 83, 84, 90, 96, 125–128, 142, 143, 144, 149, 151, 154; enthusiasm for, declines during Cold War, 113, 140, 141; enthusiasm for, declines in 1990s, 158; enthusiasm for, 1930s, 49, 50, 58, 59, 61–62; enthusiasm for, WW II, 73, 74, 91; ephemerality of, 159; fades during Cold War, 126; FDR's "four policemen" model, 81–82, 84, 96, 105, 110; 125; FDR's sheriff-posse model of, 110, 157, 160; "four policemen" model, revived in 1990s, 159–160; France supports, 26, 41, 42; future prospects of, considered, 159, 160; ignored in 1944 U.S. election, 98; Kipling's vision of, 9; Kirk proposal, 111; Korean War, 133; in Kosovo, 1999, 155; LTEP and, 10; in League Covenant, 28; McCloy-Zorin Statement, 145; NATO forces and, in Balkans, late 1990s, 154–158; New Commonwealth promotes, 1930s, 58; in 1960s, 128, 145, 148: pacifists oppose, 1920–1930s, 59; plans for proliferate, 1940s, 62; popular culture images of, 40, 75; post-Vietnam proposals, 149; post–WW II momentum, 124; religious proposals for, World War II, 75; revives in 1990s, 151–154; sidelined at Dumbarton Oaks, 105, 106–108; Theodore Roosevelt proposal, 8; U.S. attitudes toward schizoidal, 153; Wilson abandons, 1919, 23, 26–28. *See also* Bennett, Air Vice Marshal Don; Bishop, Air Marshal William; Bourgeois, Léon; China; Churchill, Prime Minister Sir Winston S.; Clark-Sohn plan; Culbertson, Ely; Dumbarton Oaks, United Nations Organizing Conference at; France; Great Britain; Herriot, Edouard; International air force concept; League of

Nations; McCloy-Zorin "Joint Statement of Agreed Principles for Disarmament Negotiations"; Paul-Boncour, Josef; Roosevelt, President Franklin D.; Roosevelt, Theodore; Spain; Trenchard, Genaral/Air Chief Marshal Lord Hugh; United Nations; United States; USSR; Wallace, Vice President Henry Agar; Willkie, Wendell
International Police Task Force (IPTF): in Balkans, 1990s, 154; ground rules set for, in Dayton peace talks, 1995, 156
Interservice rivalry, in U.S. armed forces, 85, 130, 142. *See also* "Battle of the Potomac"; "Revolt of the Admirals"; United States: Air Force; United States: Army; United States: Army Air Forces
IPF. *See* International police force concept
IPTF. *See* International Police Task Force
Iran, 125; Shah of, falls, 1979, 152
Iraq, 157; British air control in, 1920s-1930s, 136; and DESERT STORM, 154. *See also* Gulf War; Hussein, Saddam; Kuwait
"Irreconcilables," U.S. senators opposed to Versailles treaty, 22, 27
Isolationism/isolationists, U.S., 33, 35, 40, 41, 50, 51, 52, 56; defeat of, in 1944 elections, 108; rejection of, by religious groups, 75; residual power of, 1941–1944, 104, 105; revival in 1990s, 156. *See also* Hemispheric Defense; Neoisolationists
Italy, 104, Allies advancing in, 1944, 97; Allies differ over occupation, 110; as Axis power, 40; bombs Ethiopia, 47; and Corfu affair, 34; first military use of airplanes by, 19; invades Ethiopia, 40, 53; proposes banning bombers, 47; Saar police contingent of, 46; territorial claims of, at Paris Peace Conference, 25. *See also* Ethiopia; Fascism/Fascists; Mussolini, Benito

Japan, 51, 104; air power of, 55; as Axis power, 40; bombed by USAAF, 109, 112, 124; bombs China, 72; Britain breaks alliance with, 36; conquers Manchuria, 40, 47, 50, 58; invades China, 1937, 40, 47; leaves League of Nations, 40, 46, 50; rejects naval limits, 1936, 47; Shanghai bombed by, 1932, 46; versus U.S. on racial equality at Paris Peace Conference, 24; victories of, 1941–1942, 88; Washington Naval Conference and, 34, 47; Wilson opposes over Siberia and China, 26. *See also* China; World War II
Joad, Cyril, supports IPF concept, 59
Johnson, Amy, pioneer aviatrix, 43
Johnson, Senator Hiram, opposes international super state, 1943, 95; opposes ratifying League of Nations Covenant, 28;
Johnson, Secretary of Defense Louis, and interservice squabbles, late 1940s, 132
Johnson, Senator Lyndon B. (later President): picks bombing targets in Vietnam War, 147; strikes moderate pose on Vietnam, 146. *See also* Johnson (Lyndon B.) administration
Johnson (Lyndon B.) administration: cites rise of Axis as precedent in Vietnam era, 158; critiques of Vietnam policy, 148; enters Vietnam War, 145. *See also* Goldwater, Senator Barry M.; Gulf of Tonkin affair/Resolution; Johnson, Senator Lyndon B.; McNamara, Secretary of Defense Robert S.; North Vietnam; Operation LINEBACKER I; Operation LINEBACKER II; Operation ROLLING THUNDER; South

Vietnam; United States; Vietnam War
Joint Chiefs of Staff. *See* United States: Joint Chiefs of Staff
"Joint Statement of Agreed Principles for Disarmament Negotiations" (McCloy-Zorin), 145

Kellogg-Briand Peace Pact, 39, 40, 41, 61
Kenney, General George, 86; hopes to command UN air elements, 132
Key West, Joint Chiefs of Staff meeting at, 1949, 133
Khrushchev, Premier Nikita, coins "Wars of National Liberation" rubric, 1956, 138
King, Admiral of the Fleet Ernest J., 85; Anglophobia of, 101, 102; opposes IAF, opposes separate air force, 88; resists unification, 88. *See also* Combined Chiefs of Staff (U.S.–British); United States: Joint Chiefs of Staff; United States: Navy; World War II
King, Prime Minister Mackenzie, offers sovereignty surrender, 126
King-Hall, Stephen, concerns regarding IPF fragility, 102
Kipling, Rudyard, 1; IAF visions of, 9, 12
Kirchwey, Freda, supports League armed force, 75
Kirchwey, George, opposes LTEP, 13
Kirk, Grayson, proposes "Big Three" armed forces comprise UN IPF, 111
Kissinger, National Security Advisor/ Secretary of State Henry: opposes Wilsonian collective security, 156; post-Vietnam diplomacy of, 149. *See also* Paris Peace Talks; Vietnam War
Knox, Secretary of the Navy Frank: dies, 1944, 88; succeeded by Forrestal, 130; supports Anglo-American international naval police force, 78. *See also* International police force concept; United States: Navy; World War II
Korean War, 1950–1953, 34, 141, 149; air war in, 131–136, 138; nuclear diplomacy in, 142; weakens collective security, 136. *See also* MacArthur, General of the Army Douglas; Truman, Senator/ President Harry S.; United Nations
Kosovo: IPF format in bombing campaign, 155; NATO alliance air forces in, 1998, 154–158; 1990s, ethnic cleansing in, 158
Kuwait, 153; Iraq invades, 1990, 152

Labourites, Labour Party (British), 49; abandon pacifism, 57; favor Geneva Protocol, 37; link with RAF between world wars, 37, 57; support collective security, 57
Ladd, William, and LTEP formation, 12
Lake Mohonk conferences, 5
"Lasting Peace Through the Federation of Europe" (Rousseau), 4
League of Nations, 9, 10, 21–27, 33–35, 42, 52, 53, 80; Americans support, 53; Churchill quoted on revival, 102; Colombian-Peruvian involvement of, 1934, 34; compared with United Nations, 110, 123, 126, 142, 144; in Corfu affair, 1923, 34; decline of, 37, 40, 44, 46, 47, 49, 52, 61, 74; enthusiasm for, 1930s, 49, 50, 58; and Ethipia, 47, 49; expels USSR, 61; final attempts to revive, 62, 79, 102; French support strong military role of, 41; Germany leaves, 50, 52; and Gran Chaco War, 47; Japan leaves, 50; and Manchuria, 47, 49; mandates awarded by, 24; as peace forum, 53; revival of, as debaters' theme, WW II, 99; standing forces proposed, 39, 41, 45, 49, 50, 57, 61; U.S. opposes,

35; USSR joins, 45; Vilna, 1920 plebiscite, and, 34; weakened by Locarno Pact, 36, 50. *See also* Covenant of the League of Nations; Geneva, Switzerland; Paris Peace Conference, 1919; Treaty of Versailles, 1919; Wilson, President Thomas Woodrow
The League of Nations: A Practical Suggestion (Smuts), 23
League (of Nations) Council, 34
League of Nations Society, Britain, World War I, 10, 13
League to Enforce Peace, 14, 22; created in 1915, 9–10, 22; concept eroded at Paris Peace Conference, 22, 23, 26–27; critiques of, 13; grows in 1918, 12. *See also* Addams, Jane; Taft, President/Chief Justice William Howard; Wilson, President Thomas Woodrow
"League with Teeth," 26, 28, 37, 41, 47, 60–61, 73, 82, 103, 149; Churchill supports, 45; Hoover supports, 84
Leahy, Admiral of the Fleet William F.: named FDR's chief of staff, 85; opposes separate air force, 88
Leftists, Great Britain, 138; abandon pacifism, 57; condemn air control, 38; and imperialism, 36–37; oppose IAF, 1920s–1930s, 50; split over Nazi-Soviet Pact, 1939–1941, 75; support "four policemen" model, 81; support international peace enforcement system, 84
Leftists, U.S., 138; split over Nazi-Soviet Pact, 1939–1941, 75; support international peace enforcement system, 84
Lemmonier, Charles, United States of Europe plan of, 6
Liberal Party (British): calls for strong IPF, 1942, 83; proposes IAF, mid-1930s, 49
Liddell Hart, Sir Basil H., 33; cites WW I Independent Air Force as IAF prototype, 39, 59; on nuclear weapons influence on Cold War, 124
Lie, United Nations Secretary General Trygve, seeks UN standing armed force, 126, 133
Limited wars, during Cold War, 124, 128, 131, 138, 140, 141, 142. *See also* Brushfire war; Counterinsurgency (COIN); Unconventional warfare
Limited War in the Nuclear Age (Halperin), 140
Lippmann, Walter: Cold War collective security critique quoted, 136; critiques Wilson, 79; prefers regionalist postwar security system, 100, 103; proposes alliance of four major Allied powers, 79
Locarno Pact, 1925, 36, 37, 61; weakens League, 47
Lodge, Senator Henry Cabot, Jr.: foreign legion plan of, 137, 143; supports IPF concept in 1960s, 143
Lodge, Senator Henry Cabot, Sr.: identifies IPF paradox, 151; opposes ratifying League of Nations Covenant, 28; supports IPF concept, 1960s
London, 52; as imperial capital, 38; Luftwaffe bombs, 1940, 72; vulnerability of to air attack, 35, 55
Lord Allen, supports League "with teeth," 45
Lord Attlee: proposes IAF, mid-1930s, 49; supports disarmament, early 1930s, 57
Lord Beveridge, coins "superpower" term, rejects IAF model, 100
Lord Bryce, and LTEP formation, 12
Lord Byron, armed peace enforcement proposal, 12
Lord Cecil: as co-architect of League of Nations Covenant, 25; supports strong League, 41; supports weak League, 23
Lord Davies: cites WW I Independent Air Force as IAF prototype, 39; deems air power "prime factor"

in warfare, 59; promotes IPF concept, 58; "Temple of Peace" scheme of, 102
Lord Grey, proposes League of Nations, 13
Lord Londonderry, proposes internationalizing airlines, 48
Lord Lothian, British ambassador to U.S., proposes Anglo-American peace-keeping system, 1940, 74, 78
Lord Russell, favors passive resistance over IPF, 13
Lorimer, James, international army proposal of, 5
Lowell, A. Lawrence: persuades Taft to head LTEP, 9; suggests "automatic" air power peace-keeping, 12
Low-intensity conflict, airpower in, 141. *See also* Counterinsurgency (COIN); Limited wars; Unconventional warfare
LTEP. *See* League to Enforce Peace
Luce, Henry: geopolitics enthusiast, 83; proclaims "American Century," 1941, 80, 100
Luftwaffe: bombs Britain, 1940, 72, 73, 76; bombs Warsaw, 1939, 71; created by Hitler, 40, 55; fear of, 57; halts Allied bomber offensive, late 1943, 97

MacArthur, General of the Army Douglas, 85, 92; ponders obsolescence of war, 1945, 112; weakens UN command in Korean War, 134
McCloy, John J., discussion with Valerian Zorin, 1961, 145. *See also* Clark-Sohn plan
McCloy-Zorin "Joint Statement of Agreed Principles for Disarmament Negotiations," 145
McCormick, Robert, opposes U.S. membership in UN, 143
MacDonald Plan, 41, 52
Maciejewski, Casimir, proposes international troop units, 9
McNamara, Secretary of Defense Robert S.: and bombing of North Vietnam, 147; changes views on Vietnam War, 1990s, 148
Madariaga, Salvador: praises British Empire, 103; sees League Army as more moral than physical, 61
Manchuria, Japanese invasion of, 1931, 40, 47, 50, 58
Manhattan Project, 104, 125
Marshall, General of the Army George C.: denies U.S. airmen independence during World War II, 88, 90; orders IPF planning, 91, 107
Massive retaliation policy of John Foster Dulles, 136, 138; condemned as "brinkmanship," 139
Media, bombing portrayed in, 1930s, 50, 59; images of aircraft in, 86; neo-Wilsonian themes in, 1944, 99; used by air power enthusiasts, 51
"Merchants of Death" Hearings, U.S. Senate holds, 1930s, 51. *See also* Neutrality Acts
Metternich, Prince Klaus von, orchestates reaction, 4
Mexico, military leaders of, oppose cooperation with U.S. military, 157
Meyer, Cord, Jr., World Federalism proposal of, 125–126
Middle East, air policing in, 1920s-1930s, 48. *See also* Air policing; Royal Air Force (RAF)
Militarism, Wilson's aversion to, 24; in Germany, post–WW I, 27, 36
Military Staff Committee, United Nations, debates over at Dumbarton Oaks, 110; ends substantive deliberations, 126–127; revived in Gulf War, 153
"Militias" (U.S.), 156
Miller, David Hunter, as co-architect

of League of Nations Covenant, 25
Millis, Walter, 81; proposes IPF, 1963, 145
Milosevic, Slobodan, 158; rejects Rambouillet Agreement, 154
Mirage bombers, of French *force de frappe*, 138
Mitchell, General William ("Billy"), 12, 33, 37, 40, 43, 72, 85–86
Mission civilitrice. See Civilizing mission.
Mitterand, President François, offers UN French troops, 154
MLF. *See* Multinational forces
Monroe Doctrine, 21, 24, 26, 37, 38, 101
Morgenthau, Hans, on national versus international systems, 130–131
Morrison, Home Secretary Herbert, 57
Moscow Declaration, November 1943, calls for postwar "Big Three" joint peacekeeping, 96
MSC. *See* United Nations Military Staff Committee
Multinational forces: barriers to creating, 1930s, 47; in WW II, 84; proposal for under NATO, 137. *See also* League of Nations; United Nations
Munich crisis, 1938, 40, 50, 56, 61; casualty projections during, 72
Mussolini, Benito, 47, 51, 58; quoted, 33; moderate demeanor of, 40

NAFTA. *See* North American Free Trade Agreement
Nagasaki, atomic bombing of, 1945, 112, 113, 124, 132
Nash, Vernon, on air policing as imperialism, 131
National Defense Act of 1916, 11, 22
National Liberation Front, Vietnam, 146. *See also* Viet Cong
National Association of Manufacturers, oppose international super state, 95
National Peace Conference, Chicago, 1915, 9
National Security Act of 1947, 130
National Socialism, 25. *See also* Nazis
Nation in Arms (Jean Jaures), 14
NATO alliance: air forces of, in Kosovo, 1998, 154; assumes IPF role, 1990s, 158; civilian policing by, in Balkans, 154, 155; created, 1949, 130, 133; field commanders' nuclear authority, 146; in Korean War, 134, 135; Suez demonstrates "out-of-area" limits of, 137, 144. *See also* International Police Task Force
Naval aviation, 86–87; in Cold War planning, 132; in Korean War, 134
Nazi Germany, Allied bombing of, 89
Nazis, 76; air power of, 55, 72; divert resources to air defense, 97; gain power in Germany, 46; influence in Latin America feared, 52; invade Russia, 72; strike moderate pose, 1939–1940, 74; use air power to recruit, 37. *See also* Germany; Hitler, Adolf; Luftwaffe
Nazi-Soviet Non-Aggression Pact, 1939, 61; divides U.S.–British Left, 75
Nearing, Scott, proposes world federation to fill impending imperialist vacuum, 84
Neoimperialism, air power as a form of, 158
Neoisolationists (U.S.), 1990s, 156
Neutrality Acts, U.S., 1930s, 40, 51
Neutral nations: blamed for League's failure and World War II, 61, 72; in Cold War, 140; U.S. as, 1930s, 51
New Commonwealth, Britain, 1930s, 80, 94; promotes IPF and IAF, 58; studies postwar world order during World War II, 73

"New Look," Eisenhower administration defense policy, 136
"New World Order," 81; in Gulf War, 153; in 1990s, 157; hopes for crumble after World War I, 33; rubric coined, 11; in U.S. media, WW II, 99
Nichols, Beverley, favors IPF/IAF, 59
Nimitz, Admiral of the Fleet Chester W., 85, 130; supports interservice cooperation, 89
Nixon, Vice President/President Richard M.: post-Vietnam diplomacy of, 149; proposes troops for Indochina, 1954, 137; succeeds Lyndon Johnson, 1969, 148
Nobel, Alfred, views on international order, 8
Nobel Peace Prize, Theodore Roosevelt receives, 8
North Africa, Allied landings in, 1942, 89; Nazi air supremacy in, 1941, 72
North American Free Trade Agreement (NAFTA), and sovereignty, 156–157
North Atlantic, U.S.–German quasi-war in, 1941,
North Atlantic Treaty Organization. *See* NATO alliance
North Korea, 136; invades South Korea, 1950, 133, 134
North Vietnam, 146–149
Northwest Frontier, of India, air policing on, 48
Nuclear diplomacy, 1950s, 142
Nuclear Test Ban Treaty. *See* Atmospheric Test Ban Treaty, 1963
Nuclear weapons: constraints on use, 138, 140; dilemmas identified, 127, 139; effect on warfare, 112, 124; fallout from tests of, 139; in NATO, 132, 135–136; Soviet doctrine unclear, 127; use of considered at Dienbienphu, 137–138; utility uncertain, 113, 124, 139; weaken IAF/IPF concepts, 136. *See also* Atmospheric Test Ban Treaty, 1963; Cold War; Hiroshima; Korean War; Manhattan Project; Nagasaki; Nuclear diplomacy; SALT I; SALT II; Vietnam War

Oberkommando des Wehrmacht, OKW, 92
O'Connor, Raymond, on strategic effect of U.S. Senate rejection of League of Nations Covenant, 35
Office of Special Political Affairs, Cordell Hull creates, 1943,
O'Donnell, General Emmett "Rosie," 135
Oldenburg, Peter, Prince of, peace mission of, 5
One World (Willkie), 77, 100; Lippmann deems naive, 103
Operation CROSSROADS, 1946, 131
Operation DELIBERATE FORCE, 1995, intimidation of Serbs in, 155
Operation DESERT STORM, 1991, 151; air power role in, 154; approximation of IAF in, 152. *See also* Gulf War
Operation LINEBACKER I, 1972, 146; smart bombs used in, 149
Operation LINEBACKER II, 1972–73, 146; smart bombs used in, 149
Operation ROLLING THUNDER, 1965–1968, 146, 147
Operation TORCH, 1942, 92, 93
Operation VULTURE, plans for, 1954, 137
Organization for Security and Coordination in Europe (OSCE), 154
OSCE. *See* Organization for Security and Coordination in Europe

Pacifism: declines during Cold War, 128–129; in nineteenth-century U.S., 5; in 1920s and 1930s, 19, 34, 39
Pacifists, 33, 53, 56–58, 136; debate collective security utility, 56; faith of, in moral pressure and sanc-

tions, 41; oppose IPF/IAF, 39, 59; "pure" versus "hard-headed," 6, 41
Pact of Paris, 1928, 51, 54. *See also* Kellogg-Briand Peace Pact
Palestine, proposed as IAF bases, late 1930s, 58
Papacy, blocks league of nations, 1462, 2; collective security proposed by, 1939, 75
Paris Peace Conference, 1919, 14, 20–25, 45, 47, 60, 61, 80; Allies' vengefulness at, 103
Parkes, Henry Bamfield, urges Anglo-American postwar hegemony, 79
Patton, General George S., Jr., 85, 100; on postwar Anglo-American peace enforcement, 91
Paul-Boncour, Josef, IPF proposals of, 60. *See also* International police force concept; League of Nations; Paris Peace Conference
Peace arbitrations, late 19th and early 20th centuries, 6; William Jennings Bryan's treaties and, 22
"Peace Army" concept, 54
Peace Ballot, in Britain, 1935, 58
"Peace Guard of the World," mystical IPF scheme, 60
Peace schemes, variety of, 54
Peace Ship, of Henry Ford, 13
"Peaceways" advertisements, 50
Pearl Harbor, Japanese attack on, 1941, 73, 87, 88
Pearson, Lester, labels UN a "Toothless League," 1944, 108
Penn, William, plan for European Union of, 3–4
People's Republic of China (PRC). *See* China, People's Republic of
Perkins, Dexter, on causes of League's weakness, 99
Perpetual Peace [*Projet de paix Perpettuelle en Europe*] (Abbé de Saint Pierre), 4
Petain, Marshal of France Henri, calls for expanded strategic bomber force, 52
Peter of Oldenburg, Prince, proposes peace structures, 5
PGMs. *See* Precision guided munitions; Smart bombs
Philippines, insurgency in, 1950s, 141; U.S. bases in, Vietnam War, 146; U.S. leaves, 131
Philip II, unification plans of, 3
Pleven Plan, for supranational European army, 137
Podebrad, King George, of Bohemia, proposes league of nations, 2
Poland, 34, 35, 56, 61, 71, 76, 108, 110, 112
Popular culture: aviation images in, 43–44; images of IAF in, 60, 75–76
Popular Science, proposes Anglo-American naval patrol, 102
Posse model of postwar peace enforcement model, proposed by FDR, 110, 157, 160
Possony, Stefan, urges U.S. unilateral aerial peace-keeping force, 129
Pour la Societe des Nations (Bourgeois), 26
Power politics, erode collective security, WW II, 85
Progressive Party (U.S., 1948), Henry Wallace presidential candidate of, 124
Project CONTROL, USAF study of RAF air policing, 136–137
Propaganda, 52; Nazis' use of, 74; United Nations,' World War II, 78
Public opinion, British: on bombing retaliation, 73; three-fourths support IPF, 1943, 94
Public opinion, U.S.: ambivalence 1990s Balkan intervention, 156; Beards' view of, 99; 80 percent favors IPF/IAF, during WW II, 85, 91, 98, 108; enthusiasm for IPF/IAF after World War II, 130; favors isolationism in 1930s, 50; favors stronger League links, mid-1930s, 51; favors supporting

Allies, 1940–1941, 75; internationalist sentiment grows, WW II, 91, 98; in 1990s, 156; one-half of supports IPF, 1941, 73; opposes involvement in World War II, 75; post-Vietnam polls favor disengagement, 149; support of IPF/IAF fades in Cold War, 126; supports using force to keep peace, 1918, 12; supports using force to keep peace, World War II, 90, 98; three-fourths support IPF, 1943, 94

"Quarantine the aggressors" speech, by FDR, Chicago, 1937, 77

Racism, in U.S., and fears of IPF and UN armed forces, 78–79, 143–144; as Communist propaganda theme in Cold War, 138
Radford, Chairman of the Joint Chiefs of Staff Admiral Arthur, on helping France in Indochina, 1954, 137
RAF. *See* Royal Air Force
RAND Corporation, Gulf air war study of, 152
Reagan, President Ronald, 151; urges UN revitalization, 153
Regional security, 56, 83; endorsed by Clarence Streit, 91; endorsed by Sumner Welles, 90; in Nixon-Kissinger foreign policy, 150
Repington, Colonel Charles, defines air power as marginal, 12
"Revolt of the Admirals," 1949, 133
Rightists, Great Britain, support international peace enforcement system, 84
Rightists, U.S.: fear UN power increase, 1990s, 156; support international peace enforcement system, 1930s, 84
"Rimlands," of Eurasia, in Cold War, 132 , 138
ROLLING THUNDER. *See* Operation ROLLING THUNDER
Roosevelt, President Franklin D. (FDR), 84, 125; ambivalence on postwar international order, 78, 96, 105, 107; approves "destroyer deal," 1940, 75; approves Munich pact, 55; calls for bombing limits, 1939, 71; calls for 50,000 planes a year, 56, 89; comparison with Wilson, 105; conceals opposition to dictators, 55; conduct of the war criticized, 92; defines "Four Freedoms," 1941, 77; disdains League, 55; enthusiasm for air power, 56, 86; favors postwar collective security at Casablanca, 82; follows public opinion polls, 50, 74, 110; "four policemen" model of, revived in 1990s, 159–160; "four policemen" postwar peace enforcement mode of, 81–82, 84, 96, 105, 110, 125; names GOP Cabinet members, 74; "as naval enthusiast, 52, 85; neutralist rhetoric of, 74; proposes posse as postwar peace enforcement model, 110, 157, 160; "quarantine the aggressor" speech of, 55; seeks fourth term, 98; seeks third term, 1940, 74; sporadically favors IPF and IAF, 77, 90, 91; supports Hoover bomber ban plan, 49; supports Sumner Welles, 90. *See also* "Big Five"; "Big Four"; "Big Three"; Casablanca Conference; United Nations; United States; World War II
Roosevelt, Theodore (TR), 12, 155; Anglophilia of, 79; attacks Wilson, 10, 27; international naval police force proposed by in Nobel Prize speech, 8–9, 56, 78, 102, 145
Root, Elihu, and LTEP formation, 12. *See also* World War I
Rousseau, Jean-Jacques, plan for European Federation of, 4
Royal Air Force (RAF), 80, 88; air control/air policing of, in 1920s-

1930s, 38; in Battle of Britain, 72, 76; as an IAF in WW II, 76, 83; in 1990s, area bombing by, 109; V-bombers of, 138. *See also* Air policing; Bennett, Air Vice Marshal Don; British Empire; Churchill, Prime Minister Sir Winston; Desert Air Force; Gibson, Wing Commander Guy; Independent Air Force; International air force concept; Suez affair; Trenchard, General/Air Chief Marshal Lord Hugh; World War I; World War II

Royal Canadian Mounted Police, as IPF model, 151

Royal Navy: losses from Exocet missiles, 1982, 150; proposed peace patrol, 111; rivalry with RAF, 12, 42. *See also* World War II

Russia, 4, 13, 35, 78, 110; clashes with China, 1960s, 138; conducts "Open Skies" flights, 157; enters League of Nations, 41; ground forces of, and UN IPF plans, 111, 112; Nazi air supremacy in, 72; Nazis invade, 72; peace efforts of, in nineteenth century, 5; U.S. aid to, before Pearl Harbor, 74; views on IAF shift, at Dumbarton Oaks, 105, 107. *See also* Soviet Union; Stalin, Josef; USSR; World War II

Saar Police, 46

Saint Pierre, Abbé. *See* de Saint Pierre, Abbé

Salter, Sir Arthur, 49; views on France and the League, 60

San Francisco, first UN session at, 1945, 109, 110, 111, 123

Scholastic magazine, IPF plans described in, 96

Security Council, UN. *See* United Nations Security Council

Segregationists, fear IPF, 78–79, 143–144

Senate, U.S. *See* United States: Senate

Serbia: bombing by NATO, 1999, 76, 155, 157, 158; confronted by NATO alliance in Kosovo, 1998, 154

Shanghai: bombing of, 1932, 49; U.S.–British rivalry at, 1945, 102

Shotwell, James: helps draft UN Charter, 95; supports regional security treaties, 1930s, 50

SILVERPLATE, B-29 nuclear bomber force, 104

SIMG. *See* Special Informal Military Group

Slessor, Marshal of the Royal Air Force Sir John, on limits of air power, 109

Smart bombs, use in Operations LINEBACKER I and II, 1972, 149. *See also* Lavelle, General John; Precision guided munitions (PMGs)

Smuts, Field Marshal Jan Christiaan: deems British Empire as proto-world order system, 46; opposes militarily strong League of Nations, 1919, 23; prefers militarily weak international association of democracies, 1943, 84; on "spirit of Geneva," 53. *See also* British Commonwealth of Nations; British Empire; League of Nations; Paris Peace Conference

Socialist Party (U.S.), opposes UN, 83

Sohn, Louis, 128. *See also* Clark-Sohn plan; McCloy-Zorin "Joint Statement of Agreed Principles for Disarmament Negotiations"

Somalia, U.S. intervention in, 1993, 153, 155, 156

South Korea: forces in Korean War, 134; invaded by South Korea, 133, 134

South Vietnam, 147, 148

Sovereignty: attacks on, in U.S. press, 73; Churchill favors limiting, 45, 94; NAFTA, and, 157; Nazis use resentment of, in diplomacy, 74; as nemesis of international order

schemes, 2, 9, 28, 41, 45, 52, 53, 101, 106, 126, 137, 159; persisting vigor of, 1990s, 156; proposals to weaken, 75, 127; as source of war, 160; and UN Charter, 95
Soviet Union, 80, 141; bombs Finnish towns, 71; boycotts UN, 133; brandishes air power, 1920s-1930s, 42, 55; diplomacy of, 40; as empire, 131; first atomic bomb test, 1949, 131–132; invades Afghanistan, 1979, 151; invades Manchuria, 1945, 112; opposes British air policing, at Geneva, 42; opposes IAF proposals, 4; rejects Baruch plan, 124; in Spanish Civil War, 57; stymied by inferior foes, 151. *See also* Cold War; International air force concept; International police force concept; Korean War; League of Nations, Russia; Stalin, Josef; United Nations; USSR; World War II
Spaak, Paul Henri, discusses IPF with Churchill, 111
Spaatz, General of the Army Carl F., 86; foresees UN failure, 127; makes gentlemen's agreement with Eisenhower, 131
Spaight, J. M.: cites WW I Independent Air Force as IAF prototype, 39; promotes UN IPF "fire brigade," 127; on sovereignty, 9
Spain, 52, 110; air war in, 44, 47, 49, 50, 59, 72; as imperial power, 14; IPF proposals of, 25, 149
Spanish Civil War, 40, 52, 57, 96; bolsters collective security, 58; Franco wins, 61; Hitler and Mussolini intervene in, 47; images of bombing in, 49; U.S. Navy patrol in, 35. *See also Brigadistas;* France; Franco, General Francisco; Great Britain; International Brigades; Italy; Spain; USSR
Special Informal Military Group (SIMG), undermines IPF, 106–107
Special Subcommittee on International Organization. *See* United States: Department of State
"Spectrum of response" concept, Maxwell Taylor proposes, 138–140
Speer, Minister of Armaments Albert, reforms Nazi war production, 97
Spencer, Herbert, condemns armed peace enforcement, 22
Spykman, Nicholas, geopolitical theorist: on irrelevance of international security schemes, 123; predicts Eurasian "Rimlands" conflicts, 138
Stalin, Josef, 83, 84; distrusts Chiang Kai-shek, 96; fears aerial Bonapartism, 42; last survivor of "Big Three," 112; Moscow meeting with Churchill, 1942, 93; pact with Hitler divides Left, 75; promotes collective security, 50; strikes moderate pose in early 1930s, 40. *See also* Russia; Soviet Union; Teheran; Yalta Conference; USSR; World War II
Stanleyville, USAF carries Belgian paratroops to, 144
"Stealth" technology, 150; used against Serbia, 1999, 155
Stettinius, Undersecretary of State Edward M., later secretary of state, 84; briefly favors strong UN, 111; prefers regionalist postwar security system, 100, 103, 106, 107, 111; prefers UN armed force contingents, 110. *See also* Dumbarton Oaks, United Nations Organization Planning Conference at
Strategic Air Command, 132; role in nuclear war, 138
Strategic Arms Limitation Talks. *See* SALT I; SALT II
Strategic bombing: in Battle of Britain, 96; in China, 96; in the Cold War, 140; forces diversion of Nazi resources to air defense, 97; as IAF

modality in Cold War, 129; in Korean War, 134–138; in 1930s, 59; post-Vietnam, 150; post–WW II debates over, 129; in Serbia, 155–158; in Spain, 96; in Vietnam War, 146–147; in World War II, 87, 96, 97, 108, 109. *See also* AFHQ (Air Force Headquarters); Aircraft types; Arnold, General of the Army Henry H. "Hap"; Battle of Britain; Casablanca Conference, 1943; China; Churchill, Prime Minister Sir Winston; Doolittle, General James; France; Great Britain; Independent Air Force; Japan; Korean War; Luftwaffe; Mitchell, General William ("Billy"); Operation ROLLING THUNDER; Royal Air Force (RAF); Spanish Civil War; Trenchard, General/Air Chief Marshal Lord Hugh; United States: Air Force; United States: Army Air Corps; United States: Army Air Forces; Vandenberg, General Hoyt S.
Strategic Bombing Survey. *See* British Strategic Bombing Survey; United States: Strategic Bombing Survey (SBS)
Strategic situation, World War II. *See* World War II
Straus, Oscar, and LTEP formation, 12
Streit, Clarence: favors regionalism, 91; proposes Anglo-American unification, IPF, 76
A Study of War (Wright), "world peace force" proposed in, 148
Subcommittee on Security Problems. *See* United States: Department of State
Submarine warfare, 48
Suez affair, 1956, and limits of air power, 142; demonstrates NATO limits, 144
Sweden, 60; Saar police contingent of, 46
Swing, Raymond Gram, favors regional "enforcement of law," 94

Taft, Senator Robert: anti-internationalism of, 98–99; opposes international super state, 1943, 95
Taft, President/Chief Justice William Howard, 12, 98; involvement in LTEP, 9, 98; thwarts congressional peace efforts, 1910, 8; Wilson placates, 1919, 26. *See also* League to Enforce Peace
Tardieu Plan, 1930, 37, 60
Taylor, U.S. Army Chief of Staff/National Security Advisor/Chairman of the Joint Chiefs of Staff/Ambassador/General Maxwell D., 140; proposes "spectrum of response," 138–139
Technical innovations, 97
Teheran: "Big Three" conference, 1943, FDR presents "four policemen" peace enforcement model at, 96; U.S. embassy in, seized, 1979, 151
Third Reich: Allied bombing of, 109; collapse of, 1945; fighter production soars, 109. *See also* Germany; Nazis
Thomas, Norman, opposes UN, 83
Thompson, Dorothy: condemns British Empire, 79; condemns Culbertson peace plan, 94
Thompson, Sir Robert, COIN theorist, 141
Treaty of Versailles, 1919, 9, 15, 19, 33, 35; forbids German air force, 55; Hitler ignores and rejects, 40, 52; seen as unfair, 54; signed, 1919, 28
Trenchard, General/Air Chief Marshal Lord Hugh, 40, 43; builds links to Labour Party, 37, 57; commands Independent Air Force, 1918, 11; exaggerates IAF effect, 12. *See also* Independent Air Force; Royal Air Force (RAF)

Truman, Senator/President Harry S., 130, 132, 137; containment policy of, 136; deems Korean War a UN police action, 133; named vice-presidential candidate, 1944, 98; nuclear weapons and, 124, 125; threatens USSR with atomic bomb, 1946. *See also* Cold War; Korean War; World War II

Unconventional warfare, in Cold War, 128, 140. *See also* Brushfire war; Counterinsurgency (COIN); Limited wars
UNEF. *See* United Nations Emergency Force
Unification, of U.S. armed services, 88, 89, 130
Union of Soviet Socialist Republics. *See* USSR
United Nations (UN), 80, 82, 110, 130, 132; American Rightists oppose, 95, 98–99, 143; armed contingents, preferred by Stettinius, 110; armed contingents of, deployed during Cold War, 123; armed contingents of, proposed, 112, 130; "Army of Observation," proposed by George Fielding Eliot, 111; Atlantic Charter, 1941, leads to, 83; Bosnia, 154; Clinton administration ambivalence toward, 153, 157; Clinton administration sets U.S.-UN intervention guidelines, 156; compared with League of Nations, 110, 123, 126, 142, 144; constabulary forces of, Eisenhower administration supports idea of, 141; constabulary forces of, role in Balkans, 1990s, 154–157; contradictions in, WW II, 78; critiqued by N. D. White, 1990s, 157; critiques of, 99, 105, 153; decline in 1990s, 154; enthusiasm for, in WW II, 35, 91; evolves during Cold War, 146; IAF under UN auspices proposed, 129–130; involved in Gulf War, 1991, 152, 154; in Korean War, 133–135; nuclear weapons, control of by, proposed, 124; offers French troops to, 154; opposition to, in U.S., 143–144; police force proposals and "Revolution in Military Affairs," 152; religious groups support, 75; space warning patrol of, proposed, 1960, 144; standing forces of, proposals, opposition to, 45–46, 108, 109, 111, 123, 125, 126, 138, 142; in Suez affair, 142; superpower rivalry weakens, 152; Taft supports, 98; troops of, in Congo, 142; U.S. ignores in Vietnam, 145; U.S. support of, in arrears, late 1990s, 153, 156; viewed as defunct, early 1960s, 142. *See also* Boutros-Ghali, UN Secretary General Boutros; Congo; Dumbarton Oaks, United Nations Organization Planning Conference at; Hammerskjold, UN Secretary General Dag; Korean War
United Nations Charter, 111, 153, 155, 156; armed force provisions of, 111; final drafting of, 1945, 109, 123; initialing draft of, 1943, 90; McNamara comment on, 149; ratified by U.S. Senate, 112. *See also* Dumbarton Oaks, United Nations Organization Planning Conference at
United Nations Command (Korea, 1950–1953), as figurehead, 134
United Nations Department of Peacekeeping Operations (DPKO), 154
United Nations Emergency Force (UNEF), 1956: created, 142; feared as precedent, 144
United Nations General Assembly: first meeting of, 1945, 111; in Korean War, 133; proceedings hampered by Soviet bloc, late 1940s, 128; urges armed ready forces, 126
United Nations Implementation Force (IFOR), 154

United Nations Military Staff Committee: created, 123; ends substantive deliberations, 128; France offers troops, 1992, 154; revived in Gulf War, 152; weakness of, due to Cold War, 123
United Nations Organization (UNO), and Military Staff Committee, 110
United Nations Organization for Security and Coordination in Europe, created, 1995, 154
"United Nations Peace Force," in Clark-Sohn plan, 128
United Nations Protection Force, 154
United Nations Security Council, 108, 110, 112, 126, 133; debate over at Dumbarton Oaks, 109–111; France becomes member of, 101
United Nations standing armed forces, 45–46, 109; American Bar Association supports, 156; debates at Dumbarton Oaks over creating, 108. *See also* United Nations
United States. *See also* American South; Hemispheric Defense; Leftists, U.S.; Public opinion, U.S.; Rightists, U.S.
Air Force, 130, 140, 154; ascending power of, 104, 131; bears brunt of air war in Korea, 1950–1953, 134; Bomber Command of, in IAF role, 132; bombs North Vietnam, 134; carries Belgian paratroops to Stanleyville, 144; covert elements of, 140; created, 1947, 85; deploys intercontinental ballistic missiles, early 1960s, 144; Eisenhower's New Look favors, 136; Far Eastern Air Force of, 134–135; Halt Phase Strategy/Doctrine of, 154; in Gulf War, 155; studies RAF air control, 136–137; supported by Louis Johnson, 132–133; UN IAF role planned, 127, 130, 133–135, 138; war plans of, 138. *See also* Aircraft types; Lavelle, General John; Operation LINEBACKER I; Operation LINEBACKER II; Operation ROLLING THUNDER; Precision guided munitions (PMGs); Spaatz, General of the Army Carl F.; Symington, Secretary of the Air Force Stuart; Vandenberg, General Hoyt S.; Vietnam War, 1965–1975
ambivalence toward IAF/IPF, 141, 153
Armed Forces Far East, Bomber Command of, in Korean War, 134
armed forces of, post-Vietnam malaise and "hollow" period, 150
Army, 11, 88, 134, 140; clashes in, over air power, 85; covert elements of, 140; develops "Electronic Battlefield," 150; General Staff, War Plans Division of, 88; Marshall orders postwar peacekeeping planning by, 91; Pentomic Division, 132; role in Halt Phase Strategy, 154. *See also* Bradley, General of the Army Omar N.; Eisenhower, General of the Army/President Dwight D.; Korean War; MacArthur, General of the Army Douglas; Taylor, U.S. Army Chief of Staff Maxwell D.; World War I; World War II
Army Air Corps: created, 1926, 85, 86; long-range flights of, 1930s, 42; precision bombing doctrine of, 43; seeks independent air force, 51; war plans of, 87. *See also* AFHQ (Air Force Headquarters); Arnold, General of the Army Henry H. "Hap"; Baker Board; Mitchell, General William ("Billy"); World War II
Army Air Forces, 85, 88, 89, 104, 144; Air Transport Command of, 104; area bombing of Germany, 109; created, 1941, 85; 8th Air Force of, 87, 97; 15th Air Force of, 87; fire bombs Japan, 112; Middle East Air Force of, 76; planners blocked from UN force

planning by Navy, 98; plans for postwar UN IAF, 91, 113, 127; supports ground forces, WW II, 87; 20th Air Force of, 109. *See also* Aircraft types; Arnold, General of the Army Henry H. "Hap"; Doolittle, General James; Dumbarton Oaks, United Nations Organization Planning Conference at; Kenney, General George; Spaatz, General of the Army, Carl F.; Strategic bombing; United States: Navy
Asiatic Fleet of, 35, 97
blamed for League's failure, 61
Chief of Naval Operations, U.S., staff of, 88
in Cold War conflicts, 131
conducts "Open Skies" flights, 157
Congress: approves 1940 defense buildup, 75; asks Wilson to convene disarmament conference, 1915, 10; Churchill speaks to, 93; creates Atomic Energy Commission, 1946, 125; Gulf of Tonkin resolution of, 1964, 146; internationalist shift in, 95; IPF/IAF proposals and, 141; isolationism in, 44; members oppose UN, 110; 1918 elections and, 21, 23; 1910 Joint Resolution of, 8, 101; opposes intercontinental bomber, 51; Pacific basing committee of, 82; and pacifist pressure, 56; passes United Nations Participation Act, 1946, 125; and postwar security, 78; strengthens League links, 54; and United Nations, 105. *See also* Ball-Burton-Hatch-Hill Resolution; Covenant of the League of Nations; League of Nations; United States: House of Representatives; United States: Senate
Constitution: and declaring war, 51; and Versailles treaty ratification, 27, 28
Department of State: Advisory Committee on Post-War Foreign Policy, 79, 104; Commission to Study the Organization of Peace, 96; Group to Study the Organization of the Peace, 73, 79, 95; and Joint Board, 88; officials favor IAF at Dumbarton Oaks opening, 105, 107; Office of Special Political Affairs, created, 1944, 100; secret poll, 1945, 108; Special Subcommittee on International Organization, 81; Subcommittee on Security Problems, 80; supports "UN Peace Force," 128; Welles ousted from, 1944, 90. *See also* Byrnes, Justice/Secretary of State James F.; Dumbarton Oaks, United Nations Organization Planning Conference at; Hull, Secretary of State Cordell; Stettinius, Undersecretary of State Edward M.; Welles, Undersecretary of State Sumner
entry of, into World War I, 10, 11, 23
expanding power of, 14, 138
House of Representatives: Armed Services Committee of, 130. *See also* United States: Congress
as "imperial police power," 155
isolationism of, 33, 35, 35, 40, 41
Joint Board, 88
Joint Chiefs of Staff: and bombing of North Vietnam, 147; created, 1942, 85, 88; drafts IAF plans, 82; Key West meeting, 1948, 133; oppose IPF, 1943, 107; Special Informal Military Group and, 107. *See also* Anglo-American relations; Arnold, General of the Army, Henry H. "Hap"; Combined Chiefs of Staff (U.S.–British); Johnson (Lyndon B.) administration; King, Admiral of the Fleet Ernest J.; Korean War; Marshall, General of the Army George C.; Vietnam War; World War II
and League of Nations, 9, 10, 12, 14, 21–27, 33–34, 40

Marine Corps, 92; role in Halt Phase Strategy/Doctrine, 154; Union Jack raised by, on Tarawa, 79
Monroe Doctrine, 21, 24, 26, 37
National Guard, 11
Navy, 35, 88, 104, 134, 140; air power and, 85, 87; Asiatic Fleet of, 35, 97; "Battle of the Potomac," involvement in, 132, 140; in China, 1920s-1930s, 35, 125; Eisenhower's New Look benefits, 136; General Board of, 97; keeps USAAF planners out of UN force planning, 98; in Korean War, 134; opposes IAF/IPF, 85, 101; opposes separate air force, 97; opposes unification, 97, 130; secretary of, resigns, 1949, 133; Spanish Civil War, 35, 125. *See also* Combined Chiefs of Staff (U.S.–British); Johnson (Lyndon B.) administration; King, Admiral of the Fleet Ernest J.; Korean War; Operation ROLLING THUNDER; Roosevelt, President Franklin D.; Roosevelt, President Theodore; Vietnam War; World War I; World War II
neutralism of, 61, 74
Neutrality Acts of, 1930s, 40
as one of the "four policemen," 81
and Paris Peace Conference, 1919, 21–27
post-Vietnam decline of, 150
racial segregationists in, fear UN, 143, 144
Senate: constitutional role, 77–78; endorses League Covenant principles, 1935, 51; holds "Merchants of Death" Hearings, 51; ratifies UN Charter, 112; rejects League of Nations Covenant, 1919, 26–28, 50, 98, 105; strengthens links with League, mid-1930s, 51. *See also* United States: Congress
signs Atmospheric Test Ban Treaty, 139
Strategic Bombing Survey (SBS): cited in Vietnam War debates, 147–148; weakens strategic bombing rationale, 129
in Suez affair, 142
UN, support of in arrears, 153, 156
in UN armed contingent, 112
in Vietnam War. *See* Vietnam War
War Department: postwar regional security system plans of, 112; reorganization of, 1942, 90
Washington Naval Conference, 34
United States of Europe concept, 37, 61
United World Federalists, 131; peace organization designs of, 74. *See also* World Federalism
"Uniting for Peace" (UN resolution, 1950), 133
Universal Peace Congresses, 7
UNPROFOR. *See* United Nations Protection Force
Urquhart, Brian, calls for standing UN armed contingents, 156
USAAF. *See* United States: Army Air Forces
USAF. *See* United States: Air Force
USSR, 78; aids Nazis, 1939–1941, 74; blamed for League's failure and World War II, 61, 72; boycotts UN meetings, 1950, 133; expelled from League, 72; and League of Nations, 40–41, 45, 61; neutralism of, 74; as one of "four policemen," 81; rapprochement with PRC, 133; signs Atmospheric Test Ban Treaty, 139; in Suez affair, 142. *See also* Russia; Soviet Union
U.S. War Aims (Lippmann), 103

Vandenberg, Senator Arthur S., abandons isolationism, 98, 108
Vandenberg, General Hoyt S., 86; breaks Eisenhower-Spaatz agreement, 131
Veblen, Thorstein, and "new world order," 13

Vergeltungswaffen (vengeance weapons), 109
Versailles treaty. *See* Treaty of Versailles
Victory Through Air Power (de Seversky book; World War II Walt Disney film), 72, 91
Vietnam War, 1965–1975, 145–149; antiwar movement and, 147; incoherent U.S. strategy in, 146; tight rules of engagement in, 147; U.S. involvement, 131, 144, 146–149. *See also* Aircraft types; ARC-LIGHT; Gulf of Tonkin affair / Resolution; Johnson (Lyndon B.) administration; Johnson, Senator Lyndon B.; Lavelle, General John; McNamara, Secretary of Defense Robert S.; Nixon, Vice President/President Richard M.; Operation ROLLING THUNDER; Operation LINEBACKER I; Operation LINEBACKER II; Paris Peace Conference; Precision guided munitions; Smart bombs; "Yankee Station"
Vilna, 1920 plebiscite, and League of Nations, 34
von Vollenhoven, Cornelius, 8
"V" weapons, 124. *See also Vergeltungswaffen* (vengeance weapons)

Wallace, Vice President Henry Agar, 95, 155; calls for postwar IAF, 82; deems World War II a "people's revolution," 81; falls from power, 1944, 90, 108; IAF proposal of, 94; modifies collective security stance, 1948, 91; proclaims "Century of the Common Man," 1942, 80; supports IPF concept, 1960s, 103, 143
War Department, U.S. *See* United States: War Department
War plans, U.S. *See* Color Plan; Rainbow Plan; United States: Army
War Resister's League, 57
Wehrberg, Hans, IPF proposed by, 45
Wehrmacht, 71, 74, 108
Weimar Germany/Republic, 25, 36
Welles, Undersecretary of State Sumner, 44, 84, 90, 103, 108; favors IPF and IAF, 82; modifies collective security stance, 91, 111
Wells, Herbert George, 12; coins "New World Order" rubric, 11; depicts IAF-based tyranny in *The Shape of Things to Come*, 42, 60
Werner, Max, proposes proto-NATO structure, 84
Western Air Pact, Anglo-French, proposed, 1935, 52
White, N. D., critiques UN in 1990s, 157
Whitten-Brown, Arthur, first trans-Atlantic flight, 43
Willkie, Wendell, 95, 103, 106, 108; calls for "One World," 100; calls for postwar IAF, 77; politically marginalized, 1944, 90; supports IPF concept in 1960s, 143
Willson, Admiral Russell, of SIMG, 107
Wilson, Thomas Woodrow, 9, 13, 14, 23, 98; assures France, 28, 47; compared with FDR, 105; collective security ideas of, rejected by Henry Kissinger, 156; Lippmann's critique of, 79; opposes France on "League with teeth," 25, 26; as peacemaker, 19–22; proposes "self-determination," sidesteps Congress, 1915, 10; struggles with Congress, 1919, 25–28; on using force for enforcing peace, 11, 26–27. *See also* Covenant of the League of Nations; Fourteen Points; League of Nations; Paris Peace Conference; World War I
Woman's National Liberal Federation, proposes IAF, mid-1930s, 49
Woman's Peace Party, 10; and LTEP, 9
Woolf, Leonard, collective security proposals of, 10

Woolsey, Theodore Salisbury, and LTEP formation, 12
World Disarmament Conference, 1933–1937, 46–49, 51, 53, 80
World Federalism: detailed disarmament plan of, 54; movement, 83
World War I, 1, 9, 11, 12, 14, 42, 49, 73, 86, 148; after-effects of, 33, 57; Pacific campaigns, Japanese air supremacy in, 1941–1942, 72; role of aircraft in, 19; source of disillusionment, 14; strategic bombing in, 43; U.S. enters, 10, 11. *See also* Air power; Independent Air Force; Mitchell, General William ("Billy"); Trenchard, General/Air Chief Marshal Lord Hugh; World War I
World War II: American power ascends in, 104; begins, 1939, 45, 71; damages imperialism, 103; deemed "people's revolution," 81; ends suddenly, 112; postwar peace enforcement concerns during, 75; strategic situation in 1942, 88–89; strategic situation in mid-1943, 92–93; strategic situation in 1944, 109. *See also* Arnold, General of the Army Henry H. "Hap"; Battle of Britain; Battle of Germany; Bomb damage assessment; Casablanca Conference, 1943; China; Churchill, Prime Minister Sir Winston S.; Combined Chiefs of Staff (U.S.–British); Doolittle, General James; Dumbarton Oaks, United Nations Organization Planning Conference at; France; Germany; Great Britain; Hull, Secretary of State Cordell; Italy; Kenney, General George; King, Admiral of the Fleet Ernest J.; Luftwaffe; MacArthur, General of the Army Douglas; Marshall, General of the Army George C.; Operation TORCH; Roosevelt, President Franklin D.; Royal Air Force (RAF); Stalin, Josef; Stettinius, Undersecretary of State Edward M.; Strategic bombing; Teheran; United Nations; United States: Army; United States: Army Air Forces; United States: Joint Chiefs of Staff; United States: Navy; Wehrmacht; Welles, Undersecretary of State Sumner
Wright, Quincy: proposes "world peace force" in 1960s, 148; on world order concept, 129
Wriston, Henry M., cites dangers of "police mentality" in IPF, 94

Zeno, proposes world order system, 2
Zorin, Valerian, discussion with John J. McCloy, 1961, 145. *See also* Clark-Sohn plan; McCloy-Zorin "Joint Statement of Agreed Principles for Disarmament Negotiations"

ABOUT THE AUTHOR

ROGER BEAUMONT has taught history at Texas A&M University since 1974. Co-founder and former North American editor of the journal *Defense Analysis,* he has written twelve books and monographs, including *Special Operations and Elite Units, 1939–1988: A Research Guide* (Greenwood, 1988); *Joint Military Operations: A Short History* (Greenwood, 1993); *War, Chaos, and History* (Praeger, 1994); and *The Nazis' March to Chaos: The Hitler Era Through the Lenses of Chaos–Complexity Theory* (Praeger, 2000); and eighty book chapters and articles.